MODELING CARBON FLUXES, NET PRIMARY PRODUCTION
AND LIGHT UTILIZATION IN BOREAL FOREST STANDS

by

Scott J. Goetz

ISBN: 0-9658564-5-3

DISSERTATION.COM

1997

MODELING CARBON FLUXES, NET PRIMARY PRODUCTION

AND LIGHT UTILIZATION IN BOREAL FOREST STANDS

by

Scott J. Goetz

Dissertation submitted to the Faculty of the Graduate School
of The University of Maryland in partial fulfillment
of the requirements for the degree of
Doctor of Philosophy
1996

Advisory Committee:

Professor Stephen Prince (Chairman/Advisor)
Professor Samuel Goward
Professor John Townshend
Professor Irwin Forseth
Dr. Charles Walthall

ABSTRACT

Title of Dissertation: **MODELING CARBON FLUXES, NET PRIMARY PRODUCTION, AND LIGHT UTILIZATION IN BOREAL FOREST STANDS**

Scott J. Goetz, Doctor of Philosophy, 1996

Dissertation directed by: Dr. Stephen D. Prince, Professor
Department of Geography
University of Maryland

The use of satellite remote sensing for modeling net primary production (NPP) was evaluated in sixty boreal forest stands spanning a range of site conditions. The work included: (i) estimating annual phenological dynamics and photosynthetically active radiation (PAR) interception with remotely sensed spectral measurements, (ii) linking annually absorbed PAR (APAR) to measured NPP and quantifying variability in light use efficiency ($_n$), (iii) evaluating sources of variability in $_n$ via mechanistic modeling of ecophysiology and associated carbon fluxes, particularly through analyses of respiratory carbon costs in relation to assimilation gains (the R:A ratio), (iv) assessing generalization of the results through an investigation of the evidence for evolutionary convergence in $_n$, the R:A ratio and assimilation per unit APAR ($_g$).

The analyses showed that observed variability in $_n$ reflects a decoupling of PAR harvesting and utilization, primarily as a result of differences in the R:A ratio. Links between $_n$, the R:A ratio and standing above-ground biomass were related to differences the carbon (energy) costs associated with synthesis and maintenance of plant constituents, and longevity (i.e. the payback period on investment in carbon gain). Estimating the R:A ratio from above-ground biomass, in order to compensate for variability in $_n$, was found to be problematic owing primarily to covariation of R and A with the amount of respiring biomass (i.e. sapwood and foliage). The analyses also showed that the differences in carbon costs between functional types (plants with related life history traits) resulted in convergence on $_g$ rather than $_n$. Variability in $_g$ was, however, introduced by stomatal control at some stressed sites. These findings were supported by the remote sensing and simulation modeling results, and the synthesis of work related to evolutionary ecology.

The primary conclusions are that variability in light utilization in these boreal forest stands was determined largely by respiratory carbon costs, and that NPP models based on light harvesting require augmentation with terms that reflect PAR utilization. Possible methods to address these issues, and their implications for NPP modeling over large areas, are discussed.

DEDICATION

To Nadine for her patience and love,
my mother for her strength and character,
and to the memory of my father
for all he taught me.

ACKNOWLEDGMENTS

There are many people to whom I am indebted for the completion of this research. First and foremost, Steve Prince provided superb critical reviews of the dissertation and related publications, and encouraged me to pursue the evolutionary ecology that opened a great many avenues of higher learning. He fulfilled the role of mentor as well as any student could hope for. Support from Forrest Hall while I was employed at NASA allowed me to complete aspects of this research that might not otherwise have been possible. I learned much about remote sensing science under his tenure. Sam Goward, John Townshend, Irwin Forseth and Charlie Walthall all provided helpful comments on the dissertation. Jaime Nickeson and Fred Huemmrich assisted me with aspects of the phenology and canopy modeling, and helped me through the times when I thought I d never complete a PhD and hold down a full-time job. Piers Sellers showed me that it was possible to do that and a whole lot more with a sense of humor. Finally, I am indebted to Frank Davis for introducing me to forest ecology and biodiversity research and for giving me the boost I needed after leaving UCSB.

TABLE OF CONTENTS

List of Tables.. vi

List of Figures... vii

Chapter I. Introduction and Background 1
    Approaches to Primary Production Modeling 2
        Statistical and Climate Models 3
        Gap Models 4
        Ecophysiological Carbon Flux Models... 4
        Production Efficiency Models... 6
            PAR Harvesting 7
            PAR Utilization 9
        Hybrid Approaches 11
    Modeling Boreal Forest Stand NPP.. 13
    Hypotheses Tested 17
    Study Area and Data.. 18

Chapter II. Remote Sensing of Net Primary Production in Boreal Forest Stands 25
    Methodological Approach... 25
        Phenology... 25
        Fractional PAR Interception 27
        Annual PAR Interception... 28
    Results... 29
        Phenological Dynamics 29
        Fractional PAR Interception 30
        Annual PAR Interception... 34
        Utilization of Annually Intercepted PAR.. 36
        Comparison of PAR Utilization with Other Forest Stands 38
    Within-Species Variability in PAR Utilization... 40
        Variation in Stand Composition and Physical Carbon Losses 41
        Variation in Below-Ground Allocation 41
        Variation in Available Resources 42
        Variation in Respiration Demands.. 43
    Regional NPP Estimation with Constant PAR Utilization... 44
    Summary of Remote Sensing Analyses 46

Chapter III. Modeling Carbon Fluxes in Boreal Forest Stands 48
    Terrestrial Carbon Exchange (TCX) Model Description 48
    Methodological Approach... 52
        Sensitivity Analysis... 54
        Parameterization 55
        Stand Simulations.. 58
    Simulation Results... 59
        Sensitivity Analysis... 59
        Comparison of Observed and Simulated NPP... 63
        Seasonal Patterns of Assimilation and Respiration 65
    Sources of Variability in PAR Utilization... 67
        Variability of PAR Utilization in Relation to R:A Ratio 71
        Variability of R:A Ratio in Relation to Biomass 77
        Gross versus Net Carbon Yield of APAR... 80
    Summary of Simulation Analyses 82

Chapter IV. Analysis of Results in the Context of Functional Convergence 85
    Resource Constraints and Adaptive Strategies 85
    Allocation Strategies, Defenses Costs and Payback Intervals 86
    Evidence for Optimization of Resource Allocation and Use Efficiency 90
        Water.. 91
        Nitrogen.. 91
        Photosynthetically Active Radiation (PAR)... 92
        Combined Resource Use Efficiencies.. 92
        Links Between Optimization and Leaf Mass Per Unit Area.. 93
        Optimized or Compromised Resource Use? 94
    Links Between Maximization of Fitness and Carbon Gain.. 95
    Implications of Functional Convergence for Modeling Net Primary Production 96
    Summary of Functional Convergence Analyses 98

Synopsis and Conclusions 100

Appendix 103

References.. 106

LIST OF TABLES

Number .... Page

1a.   Population characteristics of lowland black spruce stands. 21

1b.   Same as Table 1a but for aspen stands 22

2.   Landsat Multispectral Scanner (MSS) imagery. 23

3a.   Goodness-of-fit statistics on spruce stand phenology models and mid-season fractional canopy PAR interception 30

3b.   Same as Table 3a but for aspen stands 31

4a.   Annual PAR interception and utilization ($_i$) in spruce stands. 34

4b.   Same as Table 4a but for aspen stands 35

5.   Sequence of routines called in TCX model. 49

6.   Input variables required and output variables simulated by the TCX model. 50

7.   Maintenance respiration coefficients and carbon allocation parameters. 55

8.   Results of TCX sensitivity analysis with respect to total tree NPP. 60

9a.   Simulated annual PAR absorption and utilization ($_i$) in spruce stands. 69

9b.   Same as Table 9a but for aspen stands 70

10.   Goodness-of-fit statistics on regression relationships between various stand variables. 73

11.   Summary statistics of variation in $_n$ and $_g$ 81

LIST OF FIGURES

Number   Page

1.   Superior National Forest study area in northeastern Minnesota. 19

2.   Flow chart of the methodological approach used for the remote sensing
        portion of the analysis. 26

3.   Greenness vegetation index response to growing degree days (phenology) fit
        with the temporal profile model. 30

4.   Overstory ipar as a function of stand Greenness for a range of overstory LAI 33

5.   Canopy model simulations of overstory ipar as a function of stand Greenness........... 33

6.   Annually intercepted PAR versus measured mid-season LAI. 36

7.   Annually intercepted PAR versus annual above-ground NPP. 37

8.   A comparison of the ranges in reported values of the dry matter yield of
        intercepted or absorbed PAR. 39

9.   Observed annual NPP compared to predicted values derived from estimated
        IPAR and $_i$ values.. 45

10.   Flow chart of the methodological approach used for the carbon flux modeling
        portion of the analysis. 53

11.   Diffuse, direct and total daily PAR amounts (MJ) as a function of fractional
        cloud cover. 54

12a.   Age dependence of aspen below-ground biomass amount. 57

12b.   Same as Figure 12a but for aspen sapwood biomass. 57

13a.   Variation in simulated total spruce production as a result of combined
        variations in LAI and foliage N content. 61

13b.   Same as Figure 13a but for aspen stands. 61

14.   Variation in simulated total tree production in relation to foliage N content,
        at constant LAI. 62

15a.   Simulated AANPP for spruce simulations with measured stand variables. 64

15b.   Same as Figure 15a but for aspen stands. 64

vii

LIST OF FIGURES

Number   Page

16a.   Seasonal dynamics of daily assimilation and maintenance respiration for
        spruce stands.. 66

16b.   Same as Figure 16a but for aspen stands... 66

17.   Simulated annually absorbed PAR and annual above-ground NPP.................... 67

18.   Comparison of simulated annual PAR absorption with different approaches................ 68

19.   Simulated variation in PAR utilization as a function of variability in the
        respiration : assimilation ratio. 72

20a.   Simulated variation in PAR utilization and respiration : assimilation ratio,
        as a function of LAI. 74

20b.   Same as Figure 20a but as a function of foliage N content. 74

21a.   Simulated variation in PAR utilization and respiration : assimilation ratio
        for generalized spruce stands. 76

21b.   Same as Figure 21a but for aspen stands.. 76

22a   Variation in the proportion of spruce assimilation not used in maintenance
        respiration, in relation to standing above-ground biomass............................... 79

22b.   Same as Figure 22a but for aspen stands.. 79

23.   Annual assimilation versus annual APAR for all spruce and aspen simulations. 80

24.   Average daily assimilation per unit APAR in spruce versus aspen stands.................... 82

25.   Generalized suites of life-history traits.. 87

26.   Foliage construction costs... 89

Chapter I. Introduction and Background

The forces that drive changes in global climate and the predicted extent of future climate change have been the focus of much research and debate (Hansen et al. 1981; Ramanathan 1988; Schneider 1989). The measured concentration of atmospheric "greenhouse gases," which trap longwave radiation emitted from the Earth s surface, have increased approximately 30% since the industrial revolution, primarily as a result of fossil fuel burning (Bacastow et al. 1985; Raynaud et al. 1993; Keeling et al. 1995). The increase has, in turn, been unambiguously linked to an observed increase in near-surface warming of the atmosphere (Thompson 1995).

The response of terrestrial vegetation to the most widely studied greenhouse gas, carbon dioxide (CO_2), introduces a possible negative feedback mechanism through which CO_2 "fertilization" and associated storage of sequestered carbon in vegetation and soils may mitigate further climate warming (LaMarche et al. 1984; Prentice 1986; Bazzaz et al. 1990; K rner and Arnone 1992). A combination of other studies, including use of simulation models and laboratory and field experiments, suggests that initial increases in vegetation carbon uptake under amplified CO_2 may quickly reach new equilibrium levels or return to previous levels owing to limitations in the availability of other resources (primarily trace nutrients) (Williams et al. 1986; Pastor and Post 1988; Kirschbaum et al. 1994; Jacoby and D Arrigo 1995).

In general, because vegetation and the atmosphere are tightly coupled, as exemplified by established links between seasonal cycles of atmospheric CO_2 and vegetation activity (Tucker et al. 1986; D Arrigo et al. 1987; Conway et al. 1994), there are likely to be significant changes in the productivity of terrestrial vegetation with further increases in atmospheric CO_2 concentration (Emanuel et al. 1985; Overpeck et al. 1991; Woodward et al. 1991). The greatest changes are predicted to occur in the forest ecosystems of the northern hemisphere, primarily the boreal forests (Pearman and Hyson 1981; Houghton 1987; Post et al. 1990). Recent work with isotope tracers in general circulation models also suggest that the largest net exchange of carbon between the terrestrial vegetation and the atmosphere occurs in the higher latitudes of the northern hemisphere (e.g., Tans et al. 1990; Ciais et al. 1995; Denning et al. 1995). This net ecosystem exchange of CO_2 is determined by net photosynthesis (the difference between CO_2 assimilation and autotrophic respiration) and soil (i.e., microbial or heterotrophic) respiration, the sum of which may dynamically alternate between a net source or sink (Sundquist 1993; Conway et al. 1994; Keeling et al. 1995). Net photosynthesis through time, or net primary production (NPP), is thus an integral component of the global carbon cycle and is important to monitor, particularly in high latitude forests.

1

As a result of its important role in global carbon dynamics, research designed to quantify terrestrial NPP has received a great deal of attention in recent years. Whereas a number of approaches to NPP estimation have been developed (reviewed in the next section) one of the most promising techniques utilizes satellite remote sensing to provide a means for monitoring vegetation light capture and utilization (Lurin et al. 1994; Prince et al. 1994). Related research has suggested that remote sensing of NPP can be simplified by taking advantage of the possibility that natural selection has resulted in a narrow range of light use efficiency ( ) among plant functional types (i.e., plants with a related suite of life history traits) (Field 1991). These developments in NPP evaluation and monitoring stimulated the work reported in this dissertation.

The objectives of the research reported herein were to test the use of satellite remote sensing techniques for inferring NPP and   in boreal forest ecosystems, to examine sources of variability in the NPP and   estimates with a mechanistic ecophysiological model, and to assess generalization of the results through arguments based in evolutionary ecology. A study site in the boreal forest of northeastern Minnesota was selected because it was well suited to addressing these objectives, particularly because it provided a wide range of vegetation productivity and biomass data, and great variability in site conditions, under similar climate. Moreover, it allowed the objectives to be addressed at a spatial scale that could be verified with surface measurements. The objectives were then tested with three hypotheses that are formalized following a review of background issues and recent advances related to the subject. Because of the interdisciplinary nature of the research, which includes components of physiological ecology, radiation physics, production biology and evolutionary ecology, a glossary of terms, definitions and their associated symbols and acronyms is provided as an appendix.

Approaches to Primary Production Modeling

Methods to model NPP over large areas range from simple correlation models to complex ecophysiological models that couple vegetation - atmosphere exchange of energy, mass and momentum. The methods of modeling NPP reviewed here include statistical extrapolation of point measurements, climate models, plant growth models, ecophysiological models, and models driven with satellite remote sensing measurements. While methods based on complex ecophysiological models have distinct advantages over statistical models, techniques using remotely sensed data are able to provide information on vegetation conditions, such as leaf area index and canopy light absorption, through time. The advantages and disadvantages of the different modeling approaches, alone and in combination, are examined. The implications of evolutionary factors that may limit the ranges of variables and introduce correlations that may simplify NPP modeling are also discussed. The review focuses primarily on production

2

modeling at the stand level, but methods to extend these to large-area primary production modeling are also discussed.

Statistical and Climate Models

Statistical relationships that associate measured stand NPP with land cover type classifications have been used to provide estimates of NPP at the global scale (Bazilevich et al. 1971; Whittaker and Likens 1973; Atjay et al. 1979; Brown and Lugo 1984). This method has many problems associated with the time and place-specific dependency of the measured values, ambiguities associated with the accuracy of land cover classifications, and errors associated with the use of very few points to represent large areas. For example, estimates of the magnitude of C storage in global vegetation have varied widely (420 - 830 Gt) depending on sampling methodology and ecosystem definitions (Moore et al. 1981; Houghton et al. 1983; Detwiler and Hall 1988). Such statistical models of NPP also offer little or no predictive or monitoring capability due to their strictly empirical nature. These methods are important to pursue in relation to potential simplification of complex mechanistic models and for interpretation of broad climate-vegetation patterns predicted by such models, but they can probably not be used to reliably estimate ecosystem productivity until a more adequate sampling network is in place.

Statistical relationships have also been used to relate climate to plant production (Lieth 1975; Rosenzweig 1968; Box 1978; Brown and Lugo 1982; Emanuel et al. 1985). This approach has been limited by the fact that it assumes homogeneous responses of vegetation to temperature and precipitation, and because it incorporates few plant physiological processes. Recently, Goward and Prince (1995) characterized several examples where climate and vegetation photosynthetic activity were poorly related, in a statistical sense, owing to lagged vegetation response to climate, disturbance frequency, and unique vegetation evolutionary adaptations (e.g., precipitation anticipation).

Statistical models of vegetation and climate associations have been largely supplanted by Soil-Vegetation-Atmosphere-Transfer (SVAT) models that incorporate feedback mechanisms between vegetation and the atmosphere by coupling plant physiological processes with climate models (e.g., Dickinson et al. 1986; Sellers et al. 1986). Such changes have been used, for example, to improve Global Circulation Models (GCMs; Sato et al. 1989; Dickinson 1995) and to predict the climatic effects of deforestation (Dickinson and Henderson-Sellers 1988; Shukla et al. 1990). The focus of this work has, however, generally been to improve climate models rather than to monitor terrestrial carbon dynamics or predict ecosystem response to changes in climate.

Gap Models

Forest succession or "gap" models offer an approach to evaluate terrestrial carbon dynamics that incorporates both plant ecology and climate variability. Such models have been developed almost exclusively for temperate forest ecosystems (Botkin et al. 1972; Shugart and West 1977; Pastor and Post 1985). Gap models consist of components that simulate disturbance, the soil regime (affected by soil physical properties), the solar regime (affected by cloudiness, latitude, and topography), and forest growth dynamics on an individual tree basis.

Although gap models incorporate plant physiological processes, they do so in a prescribed manner based on a database of life history characteristics. As such, gap models lack the ability to realistically incorporate the feedback mechanisms between the plant and the atmosphere, thus are limited in applications to different regions. Moreover, these models suffer from an inability to monitor changes in the landscape, except in a stochastic manner.

Despite these limitations, gap models are continuously being improved and recent efforts have focused on incorporating plant physiology in a more mechanistic manner (Friend et al. 1993) and in linking plant growth model components with soil process and radiative transfer models in a spatial framework (Levine et al. 1993).

Ecophysiological Carbon Flux Models

Unlike gap or climate models, ecophysiological C flux models are designed to incorporate plant ecophysiology (that is, the interrelationship between the plant s functioning and its environment) and biogeochemistry, rather than plant population characteristics or life-cycle dynamics. The strength of these models lies in simulating short-term plant physiological responses to moisture, temperature and nutrient limitations (stress), and the effect of these responses through stomatal conductance on mass fluxes (CO_2 exchange through photosynthesis and respiration) and energy fluxes (evapotranspiration, sensible and latent heat exchange).

Two examples of widely cited C flux models are reviewed here to characterize the advantages and disadvantages of this approach. The first, the Terrestrial Ecosystems Model (TEM; Raich et al. 1991) simulates C fluxes at the continental scale by calculating respiration carbon costs in conjunction with process-level relationships between photosynthesis, atmospheric CO_2 concentration, moisture availability, air temperature, nitrogen (N) availability and, to some extent, the seasonality of vegetation. This model predicts C and N fluxes and pool sizes at 0.5 x 0.5 degree cells on a monthly time step and includes coupling with an independent water balance model (Vorosmarty et al. 1989). The second, Biome-BGC (BioGeochemical Cycles; Running and Hunt 1993) was originally developed for conifer forests in the north-central United States (Forest-BGC; Running and Couglan 1988; Running and Gower 1991). Biome-BGC was designed to be particularly sensitive to leaf area index (LAI) in order to incorporate LAI

4

estimates derived from satellite spectral measurements. LAI is used in calculations of canopy radiation interception, transpiration, respiration, photosynthesis, C allocation and litterfall.

In TEM, atmospheric CO_2 concentration and solar irradiance modify gross primary production (GPP) through modification of soil moisture, stomatal control of CO_2 assimilation, and air temperature effects on photosynthetic capacity ($_c$). Nutrient uptake is based on C and N abundance and relative "root or shoot" allocation. Phenology is approximated as relative changes in photosynthetic capacity (0 to 1) based on the ratio of the current to the previous month s actual evapotranspiration (AET), and the previous month s $_c$. Assimilate is lost through growth and maintenance respiration processes, and leaf litter. Below-ground allocation of NPP is divided into fine and large root production. Soil C is lost through mineralization, and N cycling is based on soil moisture and temperature. A geographic database of environmental variables (including soil, temperature, precipitation, solar irradiance and vegetation maps) is used to add a spatial component to the model. The model is then run with variables from calibration sites while adjusting rate constants to reach equilibrium and to match measured values of NPP.

TEM is based on an understanding of vegetation physiological responses to physical and biotic variables. Nevertheless, the model requires calibration for each ecosystem with empirical relationships developed from representative field data, and must be spatially extrapolated from these data as with the statistical models described earlier. This limitation is obvious in an application of the model to estimate NPP for South America using data from 12 sites, only 3 of which were actually located on the continent (Raich et al. 1991). In addition, the model is currently limited to five soil types and seven vegetation types, based on a somewhat outdated global vegetation map (Matthews 1983). Despite these limitations, the model represents a comprehensive development of C flux modeling for use over large regions, and has good potential to be better calibrated as additional data become available. It has been applied at regional scales and has demonstrated the importance of nutrient use efficiency in NPP modeling (McGuire et al. 1992; Melillo et al. 1993).

Biome-BGC treats the forest canopy as a homogenous three-dimensional leaf of depth proportional to LAI (i.e., a "big leaf"). As such, leaf-level measurements are treated as whole canopy average responses. The model has a dual (daily and yearly) timestep; daily calculations of hydrologic balance, canopy gas exchange and C assimilate partitioning to respiration and growth are passed along to the annual component of the model, which calculates above and below-ground C partitioning, litterfall, N cycling and decomposition processes. An example of the coupling of the daily and yearly timesteps of the model is the treatment of C assimilation and respiratory losses, which are accumulated daily, differenced to provide a daily measure of net photosynthesis, and passed along to the yearly part of the model for growth allocation.

Coarse resolution Biome-BGC results have been difficult to validate but components of the original Forest-BGC model have been validated at a local scale with field measurements. For example, in tests of Forest-BGC in a Montana conifer forest, simulated values of above-ground NPP were comparable to measured values for sites (Running and Couglan 1988) and for individual trees (Korol et al. 1991). In tests of the model across a wide range of sites from Alaska to Florida, the response of annual photosynthesis to increasing LAI was quite variable between the different climates (Running and Nemani 1987). Variations in respiration estimates were large, and in some cases well outside the range of literature values, however, NPP estimates generally fell within the range of values reported in the literature for each of the test sites (with the exception of a boreal forest site in Alaska). This exception was attributed to inadequate model parameterization (i.e., higher soil N-content than would be found in such a site). Running et al. (1989) note that using LAI estimated from satellite imagery at 1 km nominal spatial resolution (with which the model is typically run) may result in significant errors being propagated through model calculations, particularly in heterogeneous landscapes. Despite these difficulties, Biome-BGC is a useful and widely applied model to simulate NPP over large areas, and it uses remote sensing data to provide a temporally and spatially specific state variable (LAI).

Production Efficiency Models

The use of remotely sensed data in NPP modeling compensates for some of the limitations inherent in ecophysiological models (e.g., spatial extent), but carries with it a new set of limitations and uncertainties. The link between satellite remote sensing data and NPP has been established through the relationship between spectral vegetation indices (SVIs) and the fraction of incident PAR intercepted (f_{ipar}) or absorbed (f_{apar}) by vegetation canopies (which will be referred to as f_{par} when a distinction is not necessary).

SVIs are derived from algebraic combinations of spectral measurements available from remote sensing. The physical basis for the observed correlation between SVIs and NPP lies in two linked but independent quantities: the relationship between light reflection and fractional PAR absorption by vegetation canopies (f_{par}), and the relationship between the amount of light absorbed and NPP (PAR utilization). In order to use remote sensing to estimate NPP it is not only necessary to characterize these two separate relationships, but also to quantify incident PAR fluxes, seasonal phenological dynamics, and the areal extent and composition of ecosystems. Each of these subjects has been the focus of a great deal of research in the past two decades.

Monteith (1972, 1977) suggested that seasonal crop production was largely determined by variability in intercepted PAR. Thus, Kumar and Monteith (1982) were able to use annually integrated f_{par} and incident PAR (PAR) to provide a measure of annual NPP. The product of f_{par}, derived from a given SVI, and PAR provides a measure of production (P) defined by Steven et al. (1983) as:

$$P = \sum_{t=1}^{N} (f_{par_t} \quad PAR_t) \tag{1}$$

where t is the time interval over a growing season (of length N), $(f_{par} \quad PAR)$ is the seasonal sum of absorbed PAR, and   is the net efficiency of conversion of PAR to biomass (PAR utilization efficiency, alternately referred to as biomass - energy quotient, light use efficiency, and dry matter yield of energy).

The seasonally integrated product of incident PAR and f_{par} results in an estimate of PAR absorption by the canopy (APAR), which is in turn related to productivity rates. Applications of this simple production efficiency model (PEM), assuming a constant value of   , have provided moderate to strong correlation with surface measurements of NPP in crops (Asrar et al. 1985; Daughtry et al. 1992), semi-arid grasslands (Tucker et al. 1983; Prince and Tucker 1986; Prince 1991a), and even at continental (Goward et al. 1985) and global scales (Potter et al. 1993; Ruimy et al. 1994; Prince and Goward 1995).

Whereas these observations demonstrate a proportionality in annual PAR absorption and NPP, there are often large uncertainties in the estimates of NPP with remote sensing. Scatter in the relationships between SVI and f_{par} and between IPAR and NPP exists, and can be attributed both to errors in the remote sensing of PAR absorption and the assumption of an invariant value of   to describe PAR utilization.

PAR Harvesting

SVIs were first used to estimate f_{par} by Kumar and Monteith (1982) using Monsi and Saeki s formulation of the Bougher-Beer law for exponential extinction of light in plant canopies:

$$f_{par} = \frac{I_{par}}{PAR} = 1 - e^{-kL} \tag{2}$$

where I_{par} is the amount of PAR absorbed by the plant canopy, k is a coefficient describing the average projection of leaves in any direction, modified by a scattering coefficient based on canopy architecture, and L is the projected leaf area index (LAI). Asrar et al. (1984) and Sellers (1985, 1987) advanced this simple formulation using two-stream approximations of canopy radiative transfer, which, in turn, were later advanced to three-dimensional models that incorporated additional canopy characteristics (e.g., spatial heterogeneity and clumping; B gu 1991; Myneni et al. 1992a). Other techniques utilize geometric-optical approaches based on mixture modeling of spectral components (Strahler and Jupp 1991; Hall et al. 1995).

Regardless of the complexity of radiative transfer modeling approaches, estimation of f_{par} from SVIs is dependent on a number of factors, including view and solar geometry, leaf display, leaf optical properties, presence of non-photosynthetic elements in the canopy, the quality of irradiance (direct versus diffuse) and background reflectance. Canopy radiative transfer simulation studies of these effects, conducted by varying the soil background, time of reflectance observations (solar angles), leaf inclination angles, and photosynthetic behavior of individual leaves, show that the relationship between the SVIs and f_{par} exhibits various degrees of non-linearity, and that the non-linearity (or the differences between linear relationships) is driven mostly by background substrate properties (Choudhury 1987; Myneni et al. 1992b; Goel and Qin 1994; Roujean and Breon 1995). A combination of these effects can result in up to 30% errors in f_{par} for a given SVI (Goward and Huemmrich 1992).

It is important to quantify the degree of non-linearity in SVI-f_{par} relationships for consideration of scaling properties of the observations. Scaling of f_{par} is simplified if the same direct relationship can be used at various spatial resolutions, thereby allowing linkages between scales. Non-linearities become problematic because the results of non-linear processes vary with the scale at which they are observed. Correlation between terms may also introduce problems if they are not independent of the scale of observation. One of the primary concerns in SVI-f_{par} relationships is the degree to which non-linearity is introduced by non-photosynthetic material in the canopy, or heterogeneity in canopy cover. It has been demonstrated in a grassland ecosystem that the relationship between SVI and f_{par} becomes more linear when PAR absorption by the live (green) matter of the canopy alone is considered (Hall et al. 1992b). It has also been suggested that spatial heterogeneity in canopy cover has little effect on this scale invariance (Sellers et al. 1992). These results suggest that SVIs respond predominantly to the photosynthetically active components of the canopy, and in turn, that f_{par} estimates from SVIs scale linearly (at least in grassland cases).

Another consideration in SVI-f_{par} relationships is the degree to which an instantaneous measurement can be used to approximate diurnal variability in f_{par} with sun angle. Studies of f_{par} in several crop species suggest that accurate estimates of f_{par} from SVIs may require frequent measurements during the day, depending on canopy architecture and leaf display (Richardson and Wiegand 1989). Simulations of instantaneous measurements of f_{par} have, however, been shown to provide a stable and near-linear estimate of diurnal f_{par} when solar zenith angle is less than 60 degrees and sensor view angle is within 40 degrees of nadir (Goward and Huemmrich 1992). Moreover, Daughtry et al. (1992) found an instantaneous measurement of f_{par} at solar noon was not significantly different from diurnally averaged values of f_{par}.

A related consideration is the degree to which the entire canopy system is represented in the instantaneous (SVI) estimate of f_{par}. Sellers et al. (1992) have demonstrated through canopy radiative transfer simulations that SVIs provide a measure of entire canopy photosynthetic capacity due to a concentration of the photosynthetic machinery of canopies (N, photosynthetic enzymes and chlorophyll) in more exposed (sunlit) leaves. This analysis is supported by observations in a lodgepole pine (*Pinus contorta*) canopy, where the sensitivity of f_{par} to solar zenith angle effects was found to vary widely between different canopy layers, but to remain relatively constant for the entire canopy system (Kimes et al. 1980).

In general, the studies reviewed here have shown that the relationship between f_{par} and a given SVI is best for a continuous canopy and becomes less reliable as the canopy becomes spatially discontinuous due to background or shadowing effects, or in the presence of a large proportion of non-photosynthetic elements. While some results suggest SVIs respond predominantly to photosynthetic elements of the canopy, and that this results in spatial scale invariance of the relationship between SVI and f_{par}, additional work is needed to quantify how SVIs are affected by non-photosynthetic PAR absorption in different canopies, particularly when this component is large.

PAR Utilization

The link between seasonal light absorption and NPP has become an important topic because of its implications for the applicability of remotely sensed APAR to the estimation of NPP. If the relationship between annual NPP and canopy light absorption is similar between different plant types in different biomes, then the task of NPP estimation with remotely sensed APAR is simplified. If, however, it varies widely, then representative values must be determined for each growth-form or biome on a case by case basis (Ruimy et al. 1994) or other techniques must be devised to characterize the observed variability.

In most remote sensing applications,   has been assumed constant between vegetation types and through various growth stages because measured values, or even reasonable approximations, were unavailable. Various measures of NPP and light use by plants suggest, however, that   ranges from about 0.3 to 3.7 g MJ^{-1} among a wide range of plant species, crop varieties, and forest stands (Prince 1991b; Ruimy et al. 1994; Landsberg et al. 1996). Simulated values derived from an ecophysiological model of plant growth driven by meteorological data encompass a similar range of values (0.4 - 3.9 g MJ^{-1}) for a variety of ecosystems in North America (Running and Hunt 1993).

9

The range in observed values of annual   is known to be affected by the radiation measure used, whether incident, intercepted or absorbed, and whether total radiation or PAR is considered. Although photosynthesis is a function of PAR absorption, it is typically easier to estimate interception through relatively straightforward measures of incident and transmitted radiation. Prince (1991b) has shown that interception and absorption are not significantly different when LAI is greater than 1.0. Other factors, including photosynthetic pathway, the measure of production used (above-ground, below-ground, or total NPP), and measurement error are involved. For example, the values of   reported in the literature are rarely whole ecosystem values, which include below-ground production and may also incorporate under-story and ground cover vegetation, herbivory, decomposition, and other carbon losses.

Known variations in   also occur through the growing season as both environmental (soil moisture, vapor pressure deficit, leaf water potential) and biotic conditions (phenological stage) change. Similarly, annually averaged values of   may vary on an inter-annual basis due to variations in climatic and edaphic conditions, particularly in species that maintain relatively invariant leaf area (i.e., non-deciduous conifers). The concept of a constant   is thus open to question.

Modeling the effects of environmental stress on NPP has demonstrated that temperature, moisture, and nutrients can limit potential (unstressed) production. For example, physiological adjustments such as stomatal dependence on temperature and vapor pressure deficit have been shown to directly affect total assimilation in the short-term (Verma et al. 1986; Baldocchi et al. 1987). These observations make it important to quantify the influence of physiological adjustments on a daily basis and their net effect on annual production estimates. Several studies have shown that the range in   can be greatly reduced by accounting for environmental physiology, respiration and other carbon losses (Potter et al. 1993; Runyon et al. 1994; Landsberg et al. 1996). The observed range in   after accounting for these factors suggests it may not be necessary to establish unique values for different vegetation types to estimate NPP with remote sensing.

A hypothesis of functional convergence in resource utilization efficiencies has been proposed by Field (1991), in which it is suggested that there is an inherent stability in   as a result of evolutionary tuning of APAR to environmental conditions. The hypothesis, which has been difficult to confirm owing to a paucity of comparable measurements at a variety of spatial scales, is discussed more thoroughly in following chapters. In the analysis of the results of the work reported in this dissertation, possible causes of variability in   , both within and between species, are explored in the context of functional convergence.

10

Hybrid Approaches

A logical step to improve production modeling by ecophysiological models and production efficiency models would be to combine the two in a unified approach that minimizes the limitations of each. For example, Maas (1988) summarized the possible means by which spectral measurements can be incorporated into crop yield models. The methods include using remote sensing inputs to: (i) set the initial conditions of a growth model, (ii) update model state variables (e.g., LAI), (iii) adjust model parameters (e.g., canopy light extinction coefficient), (iv) provide frequent observations of the driving variables in model calculations (e.g., f_{par}). The appropriate methodology depends on the variables and parameters used in the model, as well as the temporal resolution of the spectral measurements.

Field (1995) generalized this by extending the consideration of hybrid models to a global scale. He suggested that remote sensing could be used in conjunction with ecophysiological models to improve terrestrial ecosystem NPP modeling through improvements in the estimation of: (i) c, f_{par} and LAI, (ii) and stomatal conductance, (iii) environmental variables required for determination of (i) and (ii).

The linkage of remote sensing and ecophysiology provides a spatially explicit method to monitor short-term variations in photosynthetic capacity through limitations imposed by the current environmental conditions. For example, remote sensing estimates of LAI used in applications of Biome-BGC, discussed earlier, are based on an approach that updates state variables. An alternative approach to merging ecophysiological modeling and remote sensing involved assigning representative C fluxes from an ecophysiological model to regional land cover classes mapped using spectral signatures from remote sensing (Bonan 1993a). The results of this exercise are important because they suggest that NPP can be estimated over large areas based solely on estimates of LAI, vegetation type and areal extent of the different vegetation types. Each of these components is amenable to estimation with remote sensing.

These studies emphasize the utility of a combination of ecosystem C flux simulations and remote sensing techniques for production modeling. Most integrations of C flux models and remote sensing has been one way, that is, remote sensing observations have been used to augment C flux models rather than vice versa. Simplified rules from C flux models may also be used to augment the remote sensing models. For example, if variability in plant growth, from whatever source, was incorporated into the   term, remote sensing models could benefit without the full computational expense and complexity of C flux models. Landsberg (1986) and Prince (1991b) have suggested this approach can be accomplished by incorporating terms representing various components of plant physiology into an extended form of the model defined by Equation 1.

11

Using the formulations of Jarvis and Leverenz (1983) to explicitly consider plant respiratory components and the effect of multiple physiological stresses on an efficiency of net production ($_n$), Prince (1991b) proposed a revised model for use with remote sensing that has since been further extended (Prince and Goward 1995). A distinctive feature of the new production efficiency model (Glo-PEM; Equation 3) is the use of spatially comprehensive and high temporal frequency observations of both biological variables (e.g., APAR and biomass), and environmental variables (e.g., vapor pressure deficit and soil water status) to evaluate stomatal control.

The model consists of an energy harvesting component (N_t S_t) and an unstressed (potential) value of expressed in terms of gross production ($_{(g),t}^{*}$). Environmental factors that reduce the conversion of APAR into plant material ( ) reduce the value of $_{(g),t}^{*}$. Respiratory losses of carbon are represented by separate growth and maintenance respiration terms (Y_g, Y_m), and losses of biomass caused by death, grazing and decay are accounted for by d.

$$P_{(n)} = \quad _t \; _t \quad _{(g),t}^{*} (N_t S_t) \; Y_{g,t} Y_{m,t} d_t \tag{3}$$

where:

$P(n)$ = NPP (g m^{-2})

     = a functional relationship accounting for the effect of environmental variables on the unstressed value of $_{(g),t}^{*}$, owing to stomatal closure, temperature, etc.

(dimensionless)

$_{(g),t}^{*}$ = potential, gross energy fixation per unit APAR (g MJ^{-1})

$N_{(t)}$ = proportion of incident PAR absorbed by canopy (dimensionless)

$S_{(t)}$ = incident PAR (MJm^{-2})

Y_m = proportion of assimilate not used in maintenance respiration (R_m).

     $Y_m = 1 - \dfrac{R_m}{P_g}$ (dimensionless) .

     P_g = gross production (photosynthesis-photorespiration)

Y_g = efficiency of conversion of assimilate into biomass, including growth respiration R_g.

     Hence $Y_g = 1 - \dfrac{R_g}{Y_m P_g}$ (dimensionless)

d = proportion of biomass lost in death, decay and grazing (dimensionless)

Note that the original use of (Monteith 1972) is equivalent to $_t \,^*_{(g,v)} Y_{g,t} Y_{m,t} d_t$ and is now referred to as $_n$. The definition of in the form of a potential, or maximum gross rate of energy fixation ($^*_{(g)}$) is based on well-established physiological measurements of non-saturated, leaf-level quantum yields for C3 and C4 species (Collatz et al. 1991). The basis for this innovation was investigated as part of this research.

An important aspect of the extended PEM approach is that it requires only inputs that can potentially be derived from remote sensing measurements. Most of the variables needed to evaluate NPP are extracted from synergistic use of optical and thermal remotely sensed observations (Goward et al. 1994; Prince and Goward 1995).

Modeling Boreal Forest Stand NPP

Both gap and ecophysiological models have been applied to stand-level simulations of NPP and to examine the role of environmental constraints on productivity in boreal forest ecosystems (Pastor and Post 1988; Bonan 1990, 1991a, 1993b; Hunt and Running 1992b). Pastor and Post (1988) were concerned mostly with examining species composition and productivity changes under climate changes induced by a doubled atmospheric CO_2 environment. They found sites with soils of high water holding capacity experienced greater increases in productivity than sites with soils of low water retention. Changes in species composition were found to alter soil N availability, which in turn amplified the vegetation changes. As a result, there was a positive feedback between C and N cycles that was bounded by negative constraints of soil moisture availability and temperature.

Bonan (1990) compared forest growth model predictions with field measurements in different forest stands from Alaska to eastern Canada to examine the applicability of a gap model in different bioclimatic regions. Measures of 79 forest properties at the study sites, including above-ground tree biomass, basal area, density, litter fall, moss and lichen biomass, and forest floor variables (litter biomass, turnover, thickness, N-concentration, and N-mineralization) were compared with model simulations. Errors in the model estimates, defined as being outside the range of observed values, varied from 0 to 10% (for central Alaska forests) to 80% (for black spruce forests in Newfoundland). The majority of errors were associated with estimates of forest floor biomass, moss biomass and moss N.

Bonan s work provides an example of the limitations of growth models calibrated with site-specific field measurements. It also identifies important limiting factors of production in boreal forest ecosystems, which in order of generally decreasing importance were: available N, depth of thaw (in permafrost areas), growing degree days (i.e., temperature), available light, and soil moisture. The particular importance of nutrient availability to boreal forest production is supported by numerous stand-level ecological studies of black spruce (Mahendrappa and Salonius 1982; Van Cleve et al. 1983; Munson and Timmer 1990) and trembling aspen (Koerper and Richardson 1980; Van Cleve and Oliver 1982; Pastor and Bockheim 1984), including some in the boreal forests of northern Minnesota (Grigal and Arneman 1970; Nordin and Grigal 1976; Alban et al. 1978). It has also been suggested that boreal broadleaf species (short lived, fast growing) are more sensitive to nutrient limitations than boreal conifers (long lived, slow growing) owing to the life history adaptations of these different functional types (Bonan 1993a).

Simulations of C fluxes with the Biome-BGC model also provided reasonable ranges of NPP at a representative aspen and spruce stand in central Canada (Hunt and Running 1992b). The model was driven by climate data and parameterized for the two species with representative values of LAI, leaf N, respiration coefficients, and assimilate partitioning to different plant components. Incident photosynthetically active radiation (PAR) was calculated from solar geometry and daylength, reduced by an estimate of atmospheric transmission based on diurnal air temperature amplitude. Fractional canopy absorption of PAR was estimated from the Monsi-Saeki formulation of the Bougher-Beer law and LAI (Equation 2), and seasonal PAR absorption was calculated as the sum of the daily product of incident PAR and fractional canopy PAR absorption (as described earlier). The authors reported estimates of gross production from 1204 to 2700 g m^{-2} yr^{-1} for spruce and 1308 to 3317 g m^{-2} yr^{-1} for aspen. Respiration losses are reported from 536 to 1047 g m^{-2} yr^{-1} for spruce and 1008 to 2004 g m^{-2} yr^{-1} for aspen. The resulting values of NPP range from 668 to 1653 g m^{-2} yr^{-1} for spruce and 300 to 1313 g m^{-2} yr^{-1} for aspen. Simulated PAR utilization efficiencies ( ) derived from NPP and seasonal absorbed PAR were 1.03 to 1.59 g MJ^{-1} for spruce stands and 0.68 to 1.54 g MJ^{-1} for aspen stands.

A significant finding from Hunt and Running s simulation was the large difference in respiration demand for spruce (39 to 44% of gross production) versus aspen (60 to 77% of gross production). This difference resulted in lower estimates of NPP in aspen than in spruce for equivalent gross production, which is inconsistent with field observations of NPP in these species (e.g., Viereck et al. 1983; Cannell et al. 1987; Woods et al. 1991). Spruce are typically less productive than aspen due to occupation of resource-poor environments, and the associated slow nutrient recycling that results from a combination of low soil temperature, and the high lignin and low N-content in spruce leaf litter (Van Cleve et al. 1983; Pastor et al. 1987). The Biome-BGC simulated   values are affected by this discrepancy in NPP simulations between the two species.

Other stand-level C flux simulations in boreal forest have been made by Bonan (1991a) using an ecophysiological model with data from 23 intensively studied stands over a wide range of conditions in central Alaska. To avoid problems of calibrating and validating the model with

14

the same forest stands, the field data were not used to estimate required parameters for the model. Instead, the parameters that estimate photosynthesis and respiration were obtained from laboratory studies. The ranges of simulated values for photosynthetic capacity, root and microbial respiration, organic matter decomposition and tree net productivity were all found to be within the range of measured values, with the exception of net production at one low productivity black spruce stand (of the eight tested). Seasonal net ecosystem CO_2 fluxes were also calculated.

Bonan s (1991a) results showed a large drawdown of atmospheric CO_2 during the growing season, which corresponded to a two month lag in atmospheric CO_2 concentrations measured at Point Barrow, Alaska. A series of 200 Monte Carlo simulations, in which each of 27 model physiological variables were chosen at random, was used to examine the sensitivity of simulated CO_2 fluxes to variable error. Twelve variables defined the response of stomatal and mesophyll resistance to environmental factors, the remaining fifteen defined tree photosynthesis and respiration. The simulation results were narrowed down to those sets of variables that reproduced observed NPP, which resulted in only 4 sets (2% of those possible). The results of model simulations with these sets suggest that the forest stands were a net annual CO_2 sink. Moreover, NPP values of the forest stands were within the range of NPP values measured at all other boreal forest sites available at the time of publication (5 independent studies).

In addition to the stand-level results, Bonan (1993a) extended the model estimates of net canopy photosynthesis to estimate net ecosystem production for a 77 km^2 area using land cover maps derived from microwave imagery of the study area. This analysis demonstrated that knowledge of land cover type is an important variable in estimating C fluxes. Use of LAI without consideration of vegetation type led to large errors (40 - 70%) in estimates of net photosynthesis. Knowledge of species composition was shown to be relatively unimportant to C-exchange estimates, as long as a general vegetation type was considered (in this case evergreen conifer versus deciduous broadleaf). LAI and vegetation type alone accounted for 94% of the variability among forest stands in net photosynthesis. In addition, regionally averaged C fluxes were found to be sensitive to just three variables: LAI, vegetation type and area occupied by the vegetation types.

The results of this analysis suggest that LAI-driven ("big-leaf") models are likely to provide erroneous estimates of NPP unless a distinction between life forms is made, but they also demonstrate that NPP estimates at a stand-level can be extrapolated to larger areas using remotely sensed estimates of a few relatively simple variables (i.e., LAI, vegetation type and areal extent). In this way remote sensing can be used to augment ecophysiological models of NPP.

Applications of ecophysiological models at the ecosystem scale suggest that these types of models have evolved to the point that they can realistically simulate observed values when properly parameterized and judiciously applied. The most significant shortcomings of the models are the explicit and sometimes complex parameterizations required and the general lack of a spatially explicit data (such as that provided by remote sensing observations). The advantages of the models include their predictive capabilities and their mechanistic foundations that incorporate both temporally dynamic vegetation-atmosphere interactions and plant physiological responses to environmental factors. It should be noted, however, that some results suggest the models may be functionally simpler than stated. This issue is explored further in later chapters of the dissertation.

Hypotheses Tested

The foregoing review of background information raised several issues related to remote sensing of NPP. It is clear that there is a need to test the feasibility of remote sensing to reliably estimate NPP at a scale that can be validated with surface measurements, to estimate PAR utilization efficiency both within and between ecosystems, and to determine whether PAR utilization varies (and, if so, why). There is also a need to model NPP in boreal forest stands owing to their suggested role in global carbon dynamics.

Three related hypotheses were developed to address these issues. The first hypothesis is that seasonally intercepted PAR is proportional to NPP in boreal forest stands. The second hypothesis is that PAR utilization is not significantly different within stands of the same species or between different species. The third hypothesis is that respiration demands and environmental stresses either partly or wholly account for any observed variability in PAR utilization.

Testing the first hypothesis requires remotely sensed estimates of annual APAR. This was undertaken by characterizing the phenological dynamics of study plots covering a wide range of age classes and environmental conditions, and utilizing a model of phenological dynamics coupled with a canopy radiative transfer model to estimate APAR. The statistical relationship between APAR and measured values of annual above-ground NPP (that is, PAR utilization or n) was then explored.

Testing the second hypothesis required calculating values of n for a range of stands of both species being studied. Analysis of variability in n between species and among stands was carried out and its relevance to regional NPP evaluation determined.

Testing the third hypothesis required characterization of variability in respiration demands and resource-use constraints (stresses) to determine whether these accounted for observed variability in n. This was accomplished through modification and parameterization of an ecophysiological carbon flux model with measured field variables, and then using the model to explore sources of variability in n. The sensitivity of the model to driving variables was assessed, and the proportion of assimilation required for respiration processes was evaluated in relation to n. Finally, the relative importance of different components of the PEM approach was assessed in the context of potential simplifying principles based in evolutionary ecology.

Study Area and Data

The study area encompasses a 2280 km^2 portion of the North American boreal forest near its southern boundary, in the Superior National Forest (SNF) of northeast Minnesota (Figure 1). Approximately half of the area is contained within the Boundary Waters Canoe Area Wilderness, the largest wilderness preserve and boreal forest ecosystem in the conterminous United States. The area is bounded to the North by the Quetico Provincial Park in Canada. The remaining half of the area is managed by the Forest Service for multiple use, which includes timber extraction.

The area was well suited to addressing the objectives because it provided a range of site conditions that result in large variability in vegetation properties, particularly NPP and biomass density, under a comparable range of climatic conditions. Between-stand variability in vegetation properties was as large as that observed at the continental scale, which was the result of a strong covariation of vegetation production with a comparably wide range of environmental conditions originating from past glaciation (as described below). Moreover, much of the data had been collected as part of an interdisciplinary experiment, summarized by Hall et al. (1992), which allowed analyses to be conducted that would not have been possible as a individual effort. Hall et al. (1992b) also provides a detailed description of the individuals responsible for the various portions of the data collection effort. Most notably, this included Dr. Kerry Woods who directed the field crews collecting biological measurements and summarized the results, as described below and reported in Woods et al. (1991).

Daily climate data for the study site were acquired from the National Climate Data Center for two stations, one 83 km to the Northeast and one 78 km to the Northwest. Climate is characterized by cold winters (-8.1^{o} C average November-March temperature) and short cool summers (17.6^{o} C average June-August temperature). Precipitation is seasonal, with more than half of the 612 mm annual average concentrated in the period between June-September.

The vegetation of the SNF is classified as Great Lakes Boreal (Rowe 1972), and has regionally dominant stands of quaking and bigtooth aspen *(Populus tremuloides, P. granditentata)*, (broadleaf deciduous species) and black spruce *(Picea mariana)* and jack pine *(Pinus banksiana)* (needleleaf evergreen species). There is evidence of strong covariation between vegetation type, soil type and disturbance history (Grigal and Arneman 1970; Heinselman 1973; Nordin and Grigal 1976). Surficial geology is a complex of granitic bedrock, glacial till, outwash deposits, alluvium, and lacustrine sediments (Wright and Watts 1969). Upland soils are predominantly shallow and rocky over bedrock with outcrops. Lowland soils typically consist of extensive water saturated peatlands. The upland soils are generally more minerotrophic and the lowland soils generally more

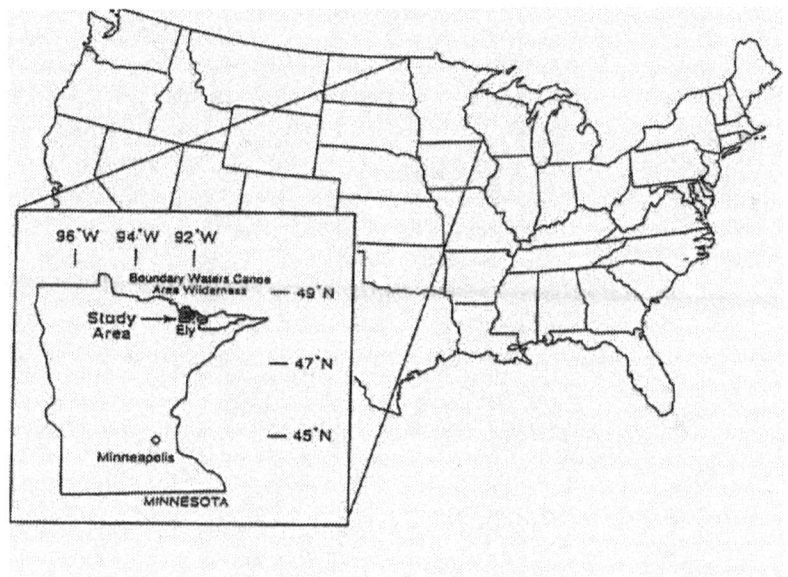

Figure 1. Study area in northeastern Minnesota.

ombrotrophic. The topography of the area is gently rolling (1200-1500 m ASL), with occasional escarpments.

Field measurements of vegetation characteristics were collected in the summers of 1983 and 1984 as part of a NASA experiment that was prematurely terminated, the results of which have been reported in Woods et al. (1991) and summarized by Hall et al. (1992a). These data are analyzed here together with additional data collected by the author, as noted below. Thirty spatially homogeneous stands each of quaking aspen and lowland black spruce, representing a range of stand density and age classes, were used to estimate LAI, annual above-ground NPP (AANPP) and a range of other stand characteristics (Woods 1988; Woods et al. 1991; Hall et al. 1992a). Each stand was sampled at five subplots within a 30 m radius of the plot center. Trees taken from outside the plots were used to develop dimensional analysis relations between diameter at breast height (DBH) and stand biological properties. Measurements of leaf expansion rates were also made at two of the aspen sites throughout the spring of 1984.

The measured biophysical characteristics of the stands are summarized in Tables 1a and 1b. In spruce stands, above-ground biomass density ranged from 700 - 15,100 g m^{-2}, LAI varied between 0.5 - 4.3, and AANPP ranged from 39 - 572 g m^{-2} yr^{-1}. In comparison, variation in above-ground biomass density among aspen stands ranged from 600 - 22,000 g m^{-2}, whereas LAI varied between 1.3 - 4.0 and AANPP ranged from 213 - 1199 g m^{-2} yr^{-1}.

Mean AANPP was estimated for each stand using a combination of allometric relationships and annual tree-ring (radial) increments for each of the previous 5 years (1979-1983) (Woods 1988). Extensive evaluation of the errors associated with the allometric relationships is provided in Woods et al. (1991). Additional discussion of the effect of errors is provided in various portions of chapter 1, particularly the section on ecophysiology models, and in the results section of the next chapter. The procedure consisted of: (1) developing predictive relationships with step-wise regression between simple stand measurements (DBH, depth of crown, tree height and basal area) and annual radial increment and terminal leader length (height growth); (2) subtracting calculated single-year radial increment and height growth from current-year radius and height; (3) calculating above-ground biomass using allometric relationships between radial increment and biomass developed from sacrificed trees for the beginning and end of the current year; (4) subtracting the two biomass calculations to get AANPP; (5) averaging AANPP over the five year period. Since the allometric relationships derived from sacrificed trees included foliar biomass, the AANPP estimates include foliar production.

Site	DBH (cm)	Biomass (g/m^2)	LAI (m^2/m^2)	NPP (g/m^2-yr)	Max AGE (years)	Density (stems/100m^2)
2	14.5	12378	2.88	324.8	146	18
12	4.5	678	0.48	39.4	101	16
14	13.2	13643	3.27	432.3	110	25
15	12.2	10680	2.69	347.6	107	22
18	4.2	1093	0.74	63.2	91	26
19	4.1	1032	0.69	58.2	91	26
38	7.3	6790	2.69	295.1	155	47
39	5.2	2373	1.32	117.6	76	38
41	13.5	11135	2.84	349.2	130	18
42	8.6	7314	2.28	258.4	119	26
45	7.3	8446	3.09	357.5	52	53
47	5.3	3527	2.00	179.9	53	58
48	9.8	9149	2.70	404.4	54	34
49	7.4	10088	3.74	409.7	69	71
50	7.6	10363	3.73	432.4	59	68
51	5.9	3620	1.69	173.9	135	39
52	9.9	10036	3.03	374.7	91	36
55	7.6	8578	3.09	360.3	74	53
56	8.1	5280	1.83	215.2	133	26
57	9.5	8293	2.60	319.1	133	32
62	4.5	894	0.59	50.5	144	20
63	3.9	1274	0.84	70.6	146	32
64	5.1	877	0.52	48.8	16	15
68	7.4	8719	3.48	382.2	101	66
100	11.0	15046	4.00	537.5	53	46
101	7.2	13500	5.42	571.8	62	107
102	5.9	7246	3.67	345.6	103	92
105	10.5	15136	4.26	537.8	80	46

Table 1a. Population characteristics of lowland black spruce (*Picea mariana*) stands. DBH is diameter at breast height, Biomass is total above-ground derived from allometric relationships, LAI is m^2 projected leaf area to m^2 surface area, NPP is net primary production, age is the maximum age of the stand, and density is the stem density of the stand. Units are as indicated.

The study site was visited by the author in the summer of 1992 to acquire better stand location coordinates, collect soil samples for physical property and nutrient analyses, and measure additional canopy reflectance characteristics. Most of the study sites were located again, and their locations were marked on aerial photographs with access descriptions. Some of the plots had been harvested. In those sites that were revisited a series of soil cores were acquired and nutrient concentrations (Mg, N, P, K), physical properties (texture, % organic matter), and chemical constituents (Ca, C and H content, Ph) were analyzed by the University of Maryland Agricultural Extension Service agronomy laboratories. C:N ratios were calculated, and the soil horizon depths were used in conjunction with texture descriptions (silt, loam, clay)

Site	DBH (cm)	Biomass (g/m^2)	LAI (m^2/m^2)	NPP $(g/m^2\text{-yr})$	Max AGE (years)	Density $(stems/100m^2)$
3	15.2	13705	2.52	563	49	9
16	16.4	13433	2.43	667	47	10
20	8.5	10340	2.46	736	64	23
21	10.1	11122	3.12	590	70	14
36	6.4	8106	2.06	538	69	24
69	2.7	2697	2.79	855	20	150
71	2.4	2250	2.85	786	18	198
72	16.1	18164	3.04	922	82	17
73	19.3	20210	3.23	812	66	11
74	17.1	21881	3.27	947	109	14
75	15.9	17648	2.94	882	65	16
79	8.8	20543	3.97	881	84	28
80	4.9	11804	2.47	624	63	37
81	13.2	17820	2.97	702	58	14
82	10.3	16171	2.71	711	98	20
83	10.1	15674	2.74	682	--	17
84	1.7	1279	2.76	529	11	311
85	12.4	15657	2.85	985	77	23
86	1.9	812	1.61	308	5	157
87	3.4	3128	2.63	999	15	105
88	3.0	3860	3.39	1199	20	161
89	2.6	2299	2.47	887	16	139
90	12.6	18350	3.20	864	69	19
92	19.9	20941	3.09	825	79	11
93	19.5	16888	2.56	686	66	9
94	1.6	726	1.65	213	9	182
95	1.6	959	2.07	258	14	226
96	13.3	16887	3.04	860	56	17
97	15.3	18785	3.32	896	76	17
98	8.1	19455	3.05	811	72	26

Table 1b. Same as Table 1a, but for quaking aspen (*Populus tremuloides*) stands

and estimates of bulk density acquired from soil series descriptions to calculate cation exchange capacity (CEC) and soil water holding capacity (SWHC). Whereas large and clear differences were noted between spruce and aspen stands (organic versus mineral soils), within-species soil samples were nearly as variable as between stands. There were also distinct differences between aspen stands on deep, fine-textured soils versus shallow, coarse-textured soils. All mature aspen stands were found on the former, whereas young aspen stands were found on both soil types.

The satellite data consisted of a total of 15 LANDSAT Multispectral Scanner (MSS) images, acquired over the study area between July 1973 and August 1990. Four scenes acquired through the 1976 growing season captured variability in early and late-season phenology, and the remainder filled out inter-annual variation in phenological response to temperature (Table 2). Each LANDSAT image was geographically referenced and radiometrically rectified to the 1983 reference image. These rectification procedures normalized the images for between-scene differences in sensor calibration and atmospheric attenuation (Hall et al. 1991b). SVI values extracted from site locations identified on low altitude aerial photographs the satellite images provided the basis for modeling phenological dynamics and annual PAR interception. Helicopter Modular Multiband Radiometer (MMR) data were also collected at intervals throughout the growing seasons of 1983 and 1984 from a height of approximately 300 m above ground level. The MMR data were not adequate to characterize seasonal phenology at most of the sites, but they were useful for validating the phenology and f_{ipar} models at some of the sites.

Platform	Day	Month	Year	D.O.Y.	GDD	SZen	SAzm
LS-1	3	July	73	184	951	32	132
LS-1	23	June	75	174	701	35	125
LS-2	21	May	76	142	203	36	131
LS-1	5	July	76	187	1115	40	117
LS-2	1	Aug	76	214	1771	39	130
LS-2	6	Sept	76	250	2511	49	139
LS-2	21	June	77	172	1027	36	122
LS-2	11	June	79	162	314	35	127
LS-3	5	June	82	157	524	34	133
LS-4	1	May	83	121	25	38	141
LS-4	18	June	83	169	477	32	133
LS-5	8	Aug	87	220	2165	40	134
LS-4	23	July	90	204	1335	36	131
LS-5	31	July	90	212	1532	38	130
LS-5	16	Aug	90	228	1886	42	134

Table 2. Landsat Multispectral Scanner (MSS) imagery used in the development of phenological response to growing degree days. Platform is the Landsat (LS) satellite number, D.O.Y. is day of year, GDD is cumulative growing degree days up to the day of the acquisition, SZen is the solar zenith angle in degrees, and SAzm is the solar azimuth angle in degrees. View angles are all at nadir. Acquisition times are all between 15:45 and 16:30 Greenwich mean time.

Daily measurements of incident total shortwave solar irradiance (300-1100 nm) made at the study area from 1972 through the 1990 provided an estimate of daily incident PAR (400-700 nm) assuming PAR was 47% of daily total shortwave (Yocum et al. 1964). Missing data were approximated using an average value derived from other years that had measurements available for that day. There were no periods during the growing season when data were missing for more than a few consecutive days in the years used for this analysis (1979 - 1984). Incident PAR was assumed the same over all stands based on the location of the pyranometer in the center of the study area. Monthly summed incident PAR measurements at the study site were compared with independent estimates of incident PAR derived from satellite observations (Dye 1993), as discussed in the next chapter.

Chapter II. Remote Sensing of Net Primary Production in Boreal Forest Stands

In this chapter high spatial-resolution remotely-sensed data are used in conjunction with a radiative transfer model to derive estimates of annual canopy PAR interception (IPAR), which are evaluated with respect to annual above-ground NPP (AANPP). Variability in PAR utilization efficiency among stands and between species is characterized, and possible sources of the observed variability are examined.

Methodological Approach

The approach taken for this component of the work, shown in Figure 2, included; (i) Characterization of growing season phenology relative to annual growing degree day using daily climate data and a series of Landsat satellite SVI images, (ii) Determination of the fraction of PAR intercepted by the canopy on a daily basis using a simple geometric-optical radiative transfer model, daily incident PAR, and daily SVI from the phenology model (iii) Calculation of IPAR from the daily values of canopy IPAR, (iv) Calculation of the utilization efficiency of IPAR for each stand ($_i$).

Note that $_i$ refers to the utilization of intercepted PAR, rather than absorbed PAR ($_a$). Both of these are synonymous with $_n$, that is, PAR utilization expressed in terms of net primary production, as described in Chapter I.

Phenology

Phenology models of vegetation response to temperature in the study area were developed using annually accumulated growing degree-days (a measure of seasonally accumulated temperature above a specified minimum value required for photosynthesis), and the Kauth-Thomas Greenness SVI (Kauth and Thomas 1976). This approach allowed a more complete characterization of growing season phenological dynamics than would have been possible with a day-of-year model by overcoming obstacles associated with poor temporal sampling frequency. This was possible because cumulative growing degree-day (GDD) is a temporal variable directly related to physiological processes associated with phenological development in aspen stands (Pauley and Perry 1954; Ahlgren 1957).

Kauth-Thomas greenness (G_i) images were calculated using all four channels of the MSS images and published greenness-transform coefficients for MSS data (Kauth et al. 1979). Stand locations were identified on low altitude aerial photographs, transferred to the MSS images, and G_i values were extracted for each individual satellite acquisition. G_i values at sites visually obscured or contaminated by cloud cover were removed from further analyses, as were

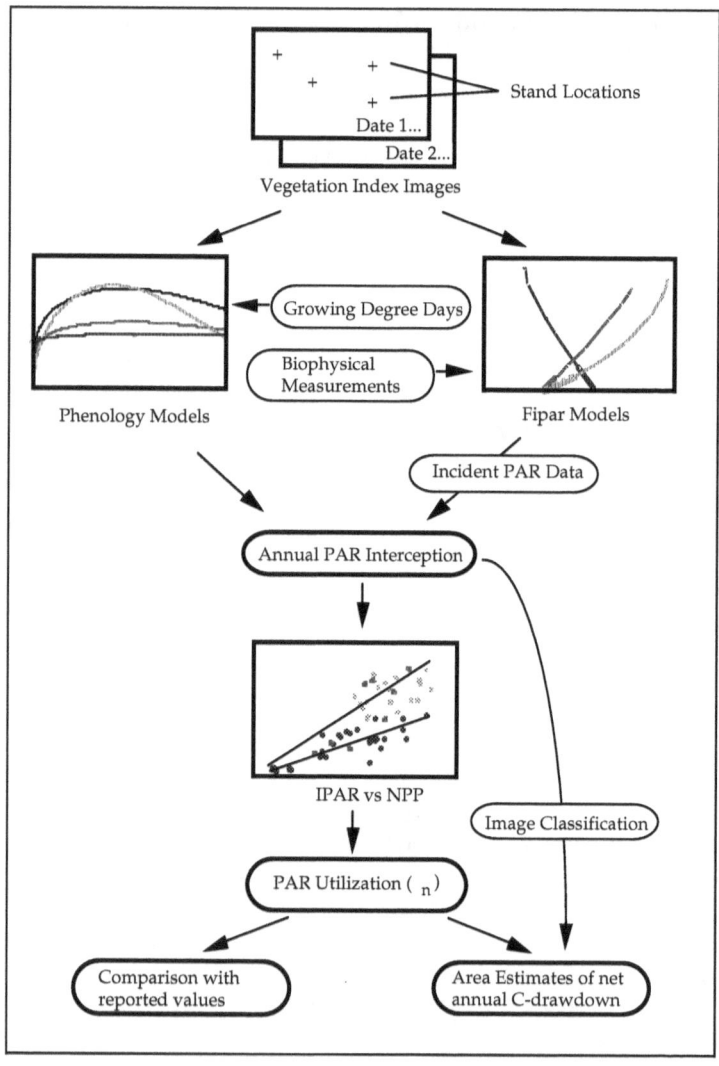

Figure 2. Flow chart of the methodological approach used for the remote sensing portion of the analysis.

values for those stands younger than the earliest acquisition dates. The "temporal profile" model of Badhwar (1984a, 1984b), which simulates phenological dynamics as a quadratic rise and exponential decay, was used to fit the G_i values at each of the stands with GDD (instead of day-of-year) as the time axis.

Growing degree-days were calculated from daily minimum and maximum temperature cycles using a sine wave model (Allen 1976) and were summed only when the average daily temperature exceeded a minimum threshold of $+5^\circ$ C. Thus the growing season was defined as starting when growing degree-days began to accumulate, and ending when they ceased to accumulate. The earliest start-date occurred on 11 April 1984, and the latest end-date occurred on 30 October 1982.

Other vegetation indices (such as normalized difference, NDVI) could be used for the phenology and f_{par} modeling. G_i was used based on its historical role in the development of the temporal profile model, and because it has been found to be a robust predictor of f_{par} over a range of conditions (Asrar et al. 1986).

Fractional PAR Interception

A simple canopy radiative transfer model (GeoSAIL, Huemmrich 1995) was used to estimate canopy reflectance and PAR interception by the stands in order to examine the relationship between annual interception and AANPP. The model was selected because it was partially developed and validated in the study area and because it utilizes a hybrid geometric-optical approach (the significance of which is discussed in the PAR harvesting section of Chapter I, and the advantages of which are shown in Chapter III relative to simple turbid media models). Two-stream equations describing reflectance and transmittance in turbid media were used in conjunction with a geometric model to estimate proportional reflectance from illuminated canopy (f_{c_i}), illuminated background (f_{bg_i}), and shadowed background (f_{bg_s}). This spectral decomposition compensates for inadequacies of a two-stream model in spatially heterogeneous canopies relative to more complex models of 3-D radiative transfer (Strahler and Jupp 1991; Hall et al. 1995).

Canopy reflectance and transmittance calculations were parameterized with measured values of LAI and leaf-level optical properties. Scene components were estimated from measured values of canopy closure, solar zenith angle, and the ratio of tree height to width. Illuminated background reflectance was specified with reflectance of leaf litter (for aspen stands) and sphagnum moss (for spruce stands), measured with a hand-held radiometer (Huemmrich 1995). Shadowed background reflectance was calculated from the measured reflectance of the illuminated background reduced by the amount of transmission through canopy. The

27

proportional contribution of scene elements was then used to calculate stand-level reflectance. Finally, fractional PAR interception (f_{ipar}) by the illuminated canopy was calculated as:

$$f_{ipar} = (1-f_{bg_i})(1-\tau_{c_i}) \qquad (4)$$

where τ_{c_i} is transmittance through the illuminated canopy in the PAR wavelengths.

The canopy reflectance component of the model was modified in the aspen stands to incorporate an additional layer representing the understory vegetation within each stand. Measured LAI and leaf optical properties of the canopy, the understory vegetation, and the background were used to parameterize each layer of the model. A range of LAI values between 0 - 5 were then used to simulate an extended range of output reflectance and f_{ipar} while canopy closure was held constant. This allowed the model to simulate a canopy filling up with leaves. Understory LAI was allowed to vary with canopy LAI based on the observed ratio between the two at mid-season.

A range of G_i values were calculated from the simulated stand-level reflectance and published greenness-transform coefficients for reflectance factor data (Crist 1985). Simulated and measured G_i values were assumed equivalent, based on the close correspondence between measured canopy reflectance from the helicopter MMR observations and the simulated values (Huemmrich 1995). The modeled relationship between simulated G_i and canopy f_{ipar} was used to obtain canopy f_{ipar} versus stand-level G_i relationships for each of the stands. This allowed estimation of PAR interception by only that component of the stand for which there were AANPP measurements - the overstory. An understory layer of the model was not required in the spruce stands as there was no significant understory component present. All simulations were run for a solar zenith angle of 32°, the angle at which the reference 1983 MSS image was acquired and to which all other scenes were rectified.

Annual PAR Interception

The fitted temporal profile models of G_i for each stand were used to estimate daily G_i. Fractional interception of PAR was then estimated from G_i with the species-specific f_{ipar} models (stand-specific f_{ipar} models in aspen). Daily canopy PAR interception was calculated as the product of f_{ipar} and PAR (Equation 1). The annual amount of canopy IPAR (IPAR) was calculated by summing the daily values over the active growing season, defined by growing degree days, for each of five years of solar measurements spanning the years for which the stand-averaged AANPP data were estimated. Note that while all MSS acquisitions (1973-1990) were used to drive the phenological response to growing degree-days, only the five years for which the AANPP data were available (1979-1983) were combined with PAR to derive annual estimates of IPAR. Utilization efficiency of IPAR (ϵ_i) was calculated on an annual basis as AANPP divided by IPAR, and averaged for the five year period.

If all intercepted PAR is assumed to be absorbed (or reduced by ~5% to account for PAR albedo), the results can be expressed as utilization efficiency of APAR (ϵ_a). The distinction between ϵ_i and ϵ_a is of no consequence to the relative comparisons that follow.

Results

Measurements of PAR at the study site through the growing season in each year of the five-year period were least in 1979 and greatest in 1980. Whereas there was large relative variation in PAR on a daily basis, there is relatively small inter-annual variation (1190 - 1350 MJ m^{-2} year^{-1}). The PAR measurements were in close agreement with estimates of PAR derived from a model driven with Total Ozone Mapping Spectrometer (TOMS) data acquired from the Nimbus-7 satellite (RMSE = 24 MJ m^{-2} month^{-1}) (Dye 1993).

Phenological Dynamics

Seasonal greenness profiles for a pair of aspen stands and a pair of spruce stands, representing different density classes, are shown in Figure 3. The phenological variation in spruce stands as depicted by variations in G_i was negligible in closed-canopy stands, whereas discontinuous stands exhibit a slight seasonality due to variations in the amount of sphagnum moss background illuminated as solar zenith angle varies through the year. In contrast, aspen stands exhibited large phenological variation in G_i, particularly during rapid leaf expansion and gradual leaf senescence.

Goodness-of-fit statistics for the phenology models are presented in Tables 3a and 3b for the spruce and aspen stands, respectively. Coefficients of determination (r^2) ranged from 0.87 to 0.99 among aspen stands ($0.001 < p < 0.01$), and 0.35 to 0.95 among spruce stands ($0.001 < p < 0.05$). Root-mean square error (RMSE) statistics ranged from 0.1 to 3.2 G_i in aspen, and 0.1 to 5.5 G_i in spruce. The low r^2 and high RMSE statistics associated with many of the spruce phenology models are a result of the small variation in G_i through the growing season, hence a low dependence of G_i on growing degree day. This is particularly true at a few spruce sites which were essentially invariant through the growing season ($r^2 = 0.18$; Table 3a). These sites were adequately characterized with a mean of the G_i values. The majority of aspen stands had RMSE values less than 1.0 G_i (24 out of 30), whereas only about half of the spruce stands met the same criterion (15 out of 29). The phenology models were found to be in good agreement with G_i values derived from the helicopter MMR observations.

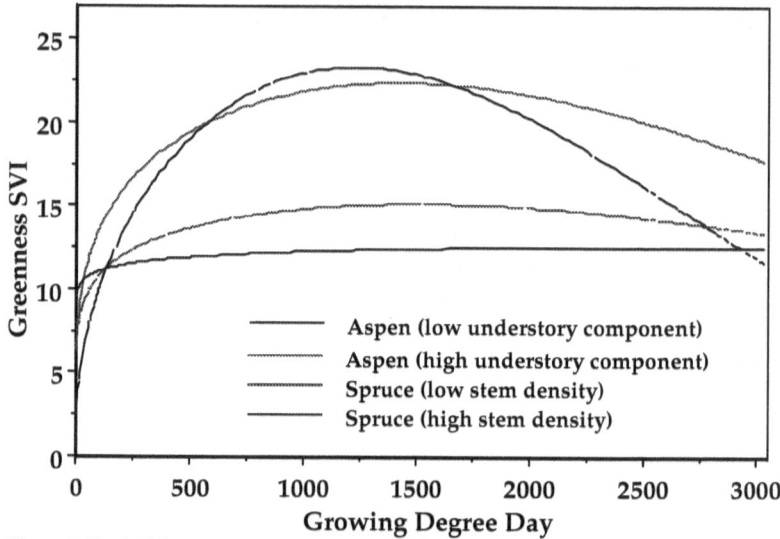

Figure 3. Typical Greenness response to growing degree days (phenology) fit with the temporal profile model, for quaking aspen with low understory component (site 36); quaking aspen with high understory component (site 87); black spruce with low density (26 stems m^{-2}; site 18); black spruce with high density (107 stems m^{-2}; site 101).

Fractional PAR Interception

The results of canopy model simulations of G_i and related values of canopy f_{ipar} are shown in Figure 4. Note again that f_{ipar} was calculated for the overstory canopy alone as AANPP data were only available for this component of the stands. Two aspen stands representing the maximum range of f_{ipar} relative to G_i are included (the same stands as depicted in Figure 3). In sites with large understory contributions to stand reflectance, canopy f_{ipar} increased less rapidly with increasing G_i than in those sites where both stand G_i and f_{ipar} were dominated by the canopy. In contrast to aspen, spruce stands exhibited an inverse relationship between G_i and f_{ipar}. In spruce, as the density grades from a closed to a discontinuous canopy, the amount of illuminated sphagnum moss increased, which acted to increase G_i. As a result, an increase in G_i reflected a decrease in canopy f_{ipar}. This result is analogous to an opaque medium obscuring a highly reflective background.

Site	r^2	RMSE (Greenness units)	$f$ipar (mid-season)
2	0.00	1.29	0.91
12	0.68	0.56	0.33
14	0.51	0.71	0.94
15	0.90	0.07	0.90
18	0.95	0.06	0.46
19	0.87	0.12	0.43
38	0.56	1.70	0.90
39	0.35	1.13	0.67
41	0.15	5.51	0.91
42	0.15	4.66	0.86
45	0.59	0.87	0.93
47	0.94	1.53	0.82
48	0.80	1.37	0.90
49	0.84	0.57	0.96
50	0.93	0.17	0.96
51	0.61	1.97	0.76
52	0.42	2.60	0.92
55	0.61	2.49	0.93
56	0.85	0.48	0.79
57	0.36	0.70	0.89
62	0.67	1.00	0.39
63	0.61	0.50	0.50
64	0.92	0.10	0.35
68	0.82	0.32	0.95
100	0.00	1.30	0.97
101	0.47	0.91	0.98
102	0.18	3.32	0.96
105	0.70	3.00	0.97

Table 3a. Goodness-of-fit statistics on spruce stand phenology (temporal profile) models, and mid-season (maximum) fractional canopy interception of incident PAR. Low r^2 values are associated with invariant G_i in closed-canopy stands.

Variation in both canopy reflectance and f_{ipar} in the spruce stands was much greater between stands than in aspen. Whereas variation in G_i in the aspen stands was driven by phenological dynamics through the growing season, variation in the spruce stands was driven primarily by between-stand differences in stem density. This result is consistent with the observation that stem density and LAI in black spruce stands co-vary, whereas aspen stands maintain nearly constant LAI despite decreasing stem-density with age (self-thinning) (Woods et al. 1991).

Site	r^2	RMSE (Greenness units)	f_{ipar} (mid-season)
3	0.95	0.59	0.68
16	0.94	0.70	0.67
20	0.93	0.92	0.69
21	0.94	1.04	0.67
36	0.92	0.67	0.56
69	0.97	0.36	0.74
71	0.87	3.24	0.75
72	0.92	0.81	0.75
73	0.96	0.36	0.78
74	0.93	0.58	0.77
75	0.96	0.55	0.75
79	0.99	0.11	0.86
80	0.98	0.15	0.70
81	0.99	0.20	0.76
82	0.98	0.19	0.72
83	0.98	0.22	0.72
84	0.98	0.50	0.74
85	0.96	0.96	0.73
86	0.68	1.05	0.50
87	0.89	1.72	0.69
88	0.92	2.66	0.80
89	0.90	1.82	0.68
90	0.93	0.60	0.72
92	0.96	0.82	0.74
93	0.96	0.52	0.69
94	0.96	0.78	0.51
95	0.99	0.14	0.59
96	0.99	0.10	0.75
97	0.96	0.49	0.79
98	0.97	0.47	0.75

Table 3b. Same as Table 3a, but for aspen stands.

Differences in LAI and f_{ipar} between aspen stands were not negligible, however, as canopy model simulations driven by measured stand variables in mid-season demonstrate. Stands with relatively low or high understory contributions to mid-season (peak) stand reflectance are indicated in Figure 5. Peak canopy f_{ipar} varied between 0.50 - 0.80 in young stands (20 year or less) and between 0.56 - 0.86 in mature stands (47 years or greater) (Table 3b). Those stands with a small understory contribution to stand reflectance were generally characterized by low G_i values relative to stands with a high understory contribution to stand reflectance. The relationships shown in Figure 4 can be concealed by such variations in stand reflectance originating from understory vegetation. Thus canopy f_{ipar} - G_i relationships must incorporate characterization of understory components in aspen stands if NPP is available only for the overstory.

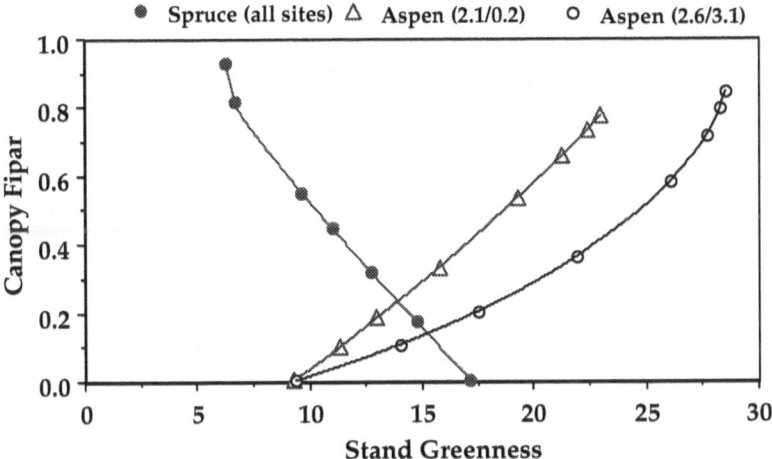

Figure 4. Overstory f_{ipar} as a function of stand Greenness for a range of overstory LAI in spruce stands, and in aspen, for different combinations of overstory (numerator) to understory (denominator) LAI values. The spruce model is generic for all sites, and the aspen model is shown for the entire range of values using the same sites as shown in Figure 3.

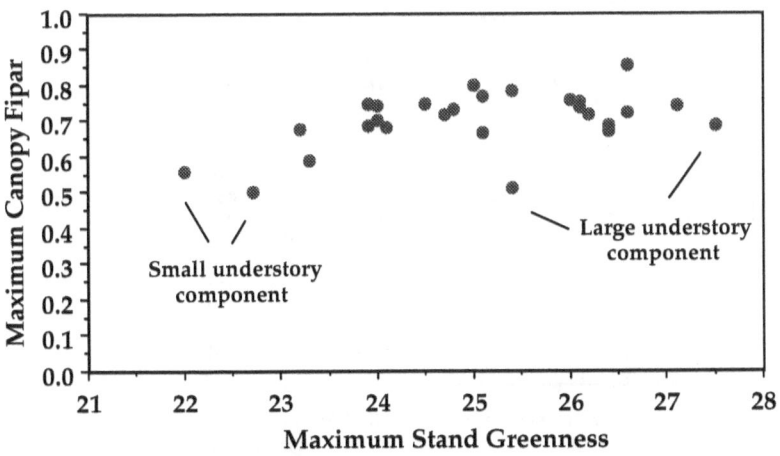

Figure 5. Canopy model simulations of overstory f_{ipar} as a function of stand Greenness at all aspen stands showing variations in Figure 4 can be concealed by understory contributions to stand greenness. The model was parameterized with measured population characteristics.

Annual PAR Interception

Much of the variation in phenological dynamics and between-stand differences in f_{ipar} are reflected in IPAR, and in annual i for spruce (Table 4a) and aspen (Table 4b). IPAR varied over a comparable range in aspen (374 - 1089 MJ m^{-2} yr^{-1}) and spruce (125 - 1057 MJ m^{-2} yr^{-1}). In initial runs of the canopy f_{ipar} model that did not incorporate the understory component of aspen stands, there was relatively little variation in IPAR. Only after the understory contribution to stand reflectance was characterized did a range of IPAR emerge, therefore only simulations incorporating an understory component were used. A similar observation was noted in the estimation of aspen LAI with remotely sensed data (Badhwar et al. 1986).

Site	IPAR (MJ/yr)	IPAR Stdev	i (g/MJ)	i Stdev
2	753.98	47.33	0.432	0.029
12	145.08	15.34	0.274	0.028
14	914.48	55.65	0.474	0.030
15	936.52	58.73	0.372	0.024
18	220.90	15.28	0.287	0.020
19	228.90	12.75	0.255	0.015
38	488.63	25.03	0.605	0.032
39	714.65	34.90	0.165	0.008
41	805.77	47.52	0.435	0.027
42	786.03	46.30	0.330	0.020
45	539.58	31.19	0.664	0.040
47	416.60	19.54	0.433	0.021
48	595.20	30.22	0.681	0.036
49	573.43	30.10	0.716	0.039
50	637.55	36.48	0.680	0.041
51	451.07	27.37	0.387	0.025
52	424.33	26.82	0.886	0.060
55	741.95	28.86	0.486	0.020
56	420.60	20.61	0.513	0.026
57	712.42	33.16	0.449	0.022
62	145.63	7.35	0.347	0.018
63	136.48	6.23	0.518	0.024
64	128.48	10.40	0.382	0.032
68	529.83	33.41	0.724	0.049
100	753.98	47.33	0.715	0.047
101	1057.25	58.38	0.542	0.030
102	765.93	30.20	0.452	0.018
105	916.15	54.35	0.589	0.037

Table 4a. Derived annually intercepted PAR (IPAR) and PAR utilization (i) for spruce stands. Units are as indicated. IPAR and i are annual averages over the five year period from 1979 through 1983. Stdev is standard deviation.

Site	IPAR (MJ/yr)	IPAR Stdev	i (g/MJ)	i Stdev
3	709	53	0.798	0.063
16	656	44	1.021	0.068
20	618	44	1.195	0.086
21	885	58	0.669	0.044
36	700	51	0.772	0.058
69	684	50	1.255	0.096
71	856	66	0.923	0.073
72	870	53	1.063	0.066
73	834	47	0.977	0.056
74	736	53	1.292	0.096
75	736	48	1.202	0.080
79	888	51	0.995	0.059
80	668	46	0.938	0.065
81	966	57	0.729	0.044
82	1004	59	0.710	0.042
83	839	48	0.815	0.048
84	804	51	0.660	0.043
85	1058	62	0.934	0.056
86	477	28	0.648	0.039
87	955	55	1.049	0.061
88	1036	61	1.161	0.069
89	729	53	1.223	0.090
90	803	48	1.079	0.065
92	765	55	1.084	0.081
93	697	47	0.988	0.068
94	374	25	0.572	0.038
95	590	43	0.439	0.033
96	1022	58	0.844	0.049
97	1014	58	0.886	0.051
98	1089	64	0.747	0.045

Table 4b. Same as Table 4a, but for aspen stands.

Canopy IPAR in both spruce and aspen stands was moderately well related to canopy LAI (r^2=0.71, 0.55 respectively at p<0.001) (Figure 6). There was, however, substantial scatter in the relationship. The variability in the LAI - IPAR relationship was attributed predominantly to errors in modeling the stand phenology and f_{ipar}, which include errors propagated from the location of stands within the images (particularly site 39, indicated in Figure 6, which was difficult to locate properly in the satellite images owing to its presence in a spectrally heterogeneous area), and from the rectification of the satellite images. Residual errors in the phenology models (see Table 3), errors in the parameterization of the f_{ipar} model with respect to canopy properties (both optical and geometric), and PAR interception by non-photosynthetic material also contribute to errors in the IPAR-LAI relationship, as do errors introduced in the estimation of LAI from field measurements. Natural factors that affect the relationship between LAI and IPAR include self-shading in the canopy (which was not characterized in the canopy reflectance model), and variations in the spatial heterogeneity of the stands (which was kept to a minimum in site selection). Comparable variability was noted in the relationship between canopy LAI and mean annual CO_2 assimilation for a range of aspen and spruce stands in Alaska (Bonan 1993a).

In spruce, IPAR also covaried with stand stem density, with a gradual increase in PAR interception as stem density increased. Others have shown that these same stands display tight linkages between stem density, LAI and the proportion of sunlit canopy within a scene (Hall et al. 1995). In contrast, the self-thinning nature of aspen stands resulted in variations in canopy PAR interception that were largely unrelated to variations in stem density.

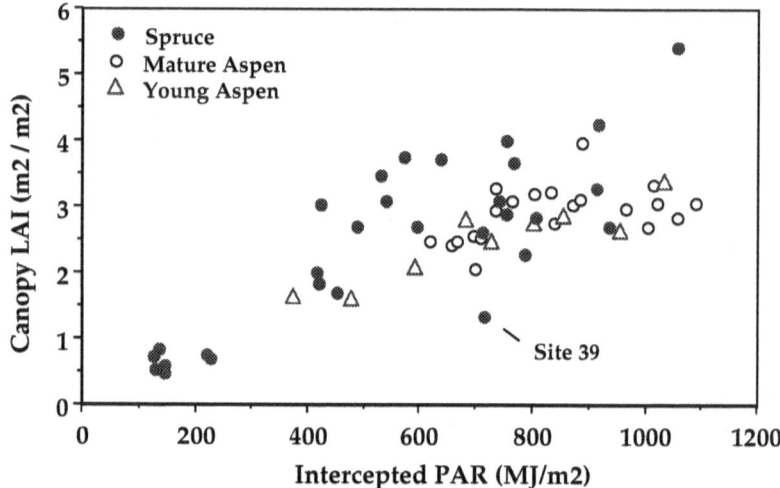

Figure 6. Annually intercepted PAR versus measured mid-season LAI. Site 39 is discussed in text.

Utilization of Annually Intercepted PAR

In examining the relationship between IPAR and AANPP for all stands (Figure 7), it is evident that the variation in NPP relative to a given value of IPAR was greater in aspen than in spruce, particularly in the mature aspen stands. Simple linear regression of IPAR on NPP for all stands provided some predictive capability based on IPAR alone ($r^2 = 0.59$, p<0.001), but the residuals were large (RMSE = 195 g m^{-2} yr^{-1}). In spruce stands, IPAR was a moderately good predictor of AANPP ($r^2 = 0.69$, p<0.001), and RMSE was reduced to 93 g m^{-2} yr^{-1}. Elimination of a single "outlier" spruce stand (site 39, highlighted in Figure 7) increased the explained variance to 76% and reduced the RMSE to 81 g m^{-2} yr^{-1}. Site 39 was just one of several such stands, but was known to be small and surrounded by dissimilar vegetation types. In aspen stands, IPAR was a relatively poor predictor of AANPP ($r^2 = 0.49$, $p<0.001$) and residuals were large (RMSE = 154 g m^{-2} yr^{-1}). Stratification of the aspen into young versus mature stands improved the relationship in young stands ($r^2 = 0.77$, $p<0.002$) but reduced it in the mature stands ($r^2 = 0.18$, $p>0.05$). Mean residuals were 182 and 123 g m^{-2} yr^{-1}, respectively.

Relatively large variability in IPAR in relation to AANPP strongly suggests that factors other than light are limiting production. This was most apparent in the mature aspen stands, and may be due in part to the fact that AANPP in mature aspen was confined to a smaller range of values (538 - 985 g m^{-2} yr^{-1}) than the young stands (190 - 1199 g m^{-2} yr^{-1}). Elimination of less productive stands through competitive replacement during stand succession may explain this observation. The relationship between IPAR and AANPP in mature aspen is discussed later.

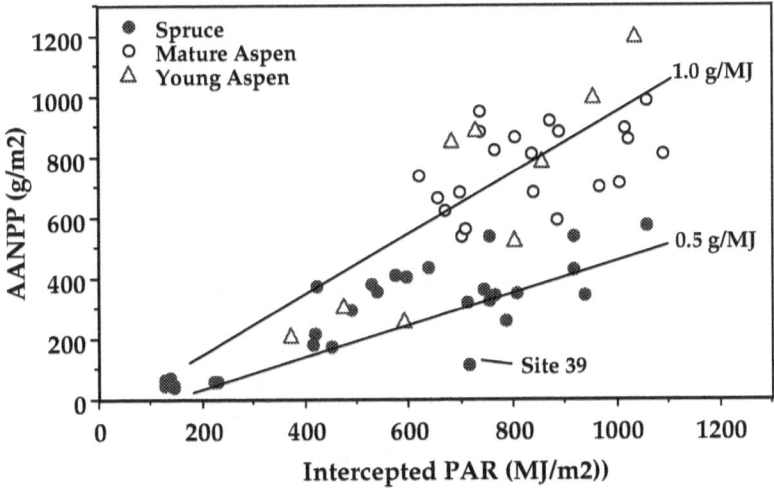

Figure 7. Annually intercepted PAR versus annual above-ground NPP. Lines indicate dry matter yield of intercepted PAR at 0.5 and 1.0 g MJ^{-1}. Site 39 is discussed in text.

Total above-ground utilization efficiency of intercepted PAR ($_i$) varied considerably among both spruce and aspen stands (Tables 4a,b). The range in $_i$ for aspen (0.44 - 1.29 g MJ^{-1}) was fully expressed over the young stands alone. This was comparable to, although higher than, the range of $_i$ in spruce (0.17 - 0.89 g MJ^{-1}). The average values of $_i$ were 0.92 (±0.22)

in aspen and 0.49 ($\pm$ 0.17) g MJ^{-1} in spruce. Differences in ε_i between the two species may reflect differences in life history, particularly differences in the energy requirements associated with different resource allocation strategies (Loehle 1988; Lambers and Poorter 1992), as discussed later.

Comparison of PAR Utilization with Other Forest Stands

Simulation studies suggest that interactions between carbon and nitrogen dynamics are related through climate variables such that temperature and moisture variations result in increasing nitrogen limitation of productivity from tropical to boreal to tundra ecosystems (McGuire et al. 1992; Melillo et al. 1993). To determine if this interaction between climate, nutrient availability and production is reflected in ε, the values of ε were compared with those for other forest stands for which such data were available. In comparing the data, it is important to note the type of radiation measure and NPP values reported. The symbol ε was used when a distinction between the utilization efficiency of intercepted PAR (ε_i) and the utilization efficiency of absorbed PAR (ε_a) could not be made or was unnecessary. Figure 8 illustrates graphically the data discussed below. All data are reported in g dry matter per MJ PAR intercepted or absorbed, as indicated, and are discussed in relation to the radiation and production measures used.

With few exceptions, most of the reported values of ε for forest ecosystems are based on light absorption as estimated from LAI and the Monsi-Saeki formulation of the Lambert-Beer law (Equation 2). Using this approach, Linder (1985) reported data for a variety of pine stands (*Pinus species*), eucalypts (*Eucalyptus globulus*), and white spruce (*Picea abies*) from which above-ground ε_a values between 0.2 - 1.5 g MJ^{-1} were calculated; Landsberg and Wright (1989) reported above-ground ε_a values between 1.16 - 1.4 g MJ^{-1} for cultivated stands of young *Populus* clones in Pennsylvania and Wisconsin; Rauner (1976) reported a gradual decrease in above-ground values of ε_a for an age sequence of oak (*Quercus robur*) on the forest-steppe of central Russia (0.24 - 1.02 g MJ^{-1}); Saldarriaga and Luxmoore (1991) reported above-ground ε_a values ranging from negligible to 0.31 g MJ^{-1}, and ε_a values for total production between 0.25 - 0.50 g MJ^{-1} for an age sequence of tropical forest stands in Amazonia, with a marked decrease related to stand age; Saldarriaga and Luxmoore (1991) also calculated ε_a values of total production between 0.2 - 1.0 g MJ^{-1} for a wide variety of forest stands studied under the International Biosphere Program (IBP) program using data reported by Jordan (1971).

It was unclear from Linder s discussion exactly how the ε values were derived. I was interested in the range of values observed, and used the data plotted in his figure 4 to calculate this.

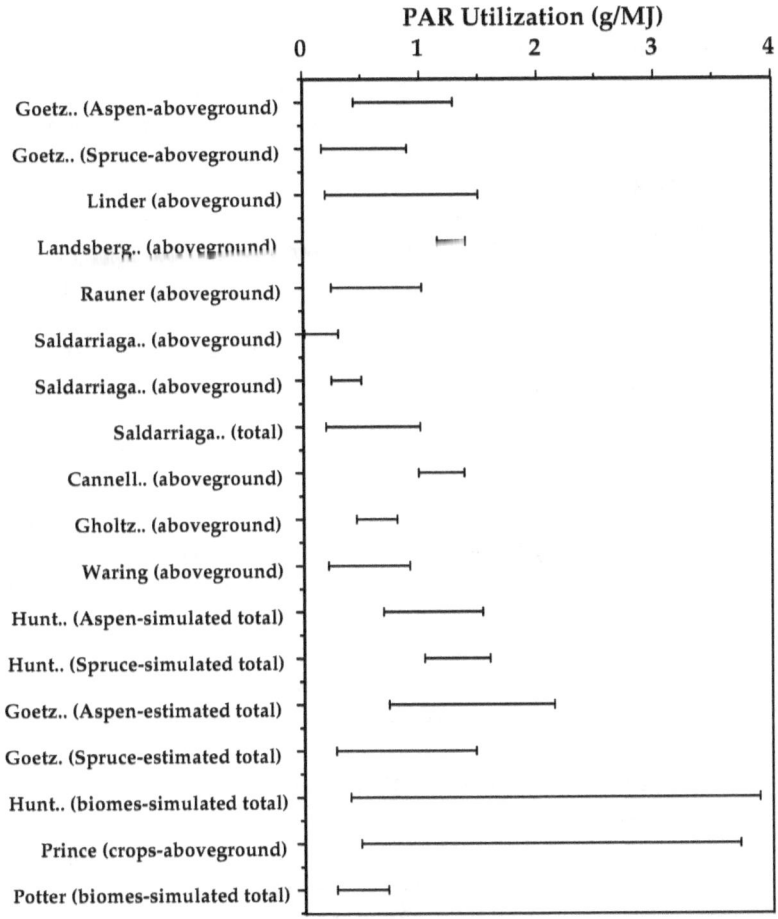

Figure 8. A comparison of the ranges in reported values of PAR utilization. See main text for more complete information on the vegetation types, the radiation and production measures used, and complete references.

In studies based on measured canopy interception of PAR and above-ground NPP, Cannell et al. (1987) reported above-ground $_i$ values of 0.99 - 1.38 g MJ^{-1} for coppiced and newly planted *Salix* stands in Scotland; Gholtz et al. (1991) reported above-ground $_i$ values of 0.46 to 0.80 g MJ^{-1} for mature slash pine (*Pinus elliottii*) forests in Florida; and Runyon et al. (1994) reported values for Sitka spruce (*Picea sitchensis*), Douglas-fir (*Tsuga heterophylla*), Ponderosa pine (*Pinus ponderosa*) and Juniper (*Juniperous occidentalis*) stands in Oregon between 0.18 - 0.92 g MJ^{-1}.

The estimates of   reported thus far were comparable to the range of values observed, although they extended over a slightly larger range (0.07 - 1.4 g MJ^{-1}). The Biome-BGC model, which uses the Monsi-Saeki formulation to estimate PAR absorption (Running and Hunt 1993), has also been used to estimate values of $_a$. Hunt and Running (1992a) report simulated values of $_a$ for total production (i.e., both above- and below-ground) from 0.4 - 3.9 g m^{-2} yr^{-1} over a range of vegetation types in North America. The observed annual above-ground values of   in a variety of agricultural systems is 0.95 to 3.72 g MJ^{-1}.

Hunt and Running (1992b) also report simulated values of $_a$ for aspen (0.68 - 1.54 g MJ^{-1}) and spruce (1.03 - 1.59 g MJ^{-1}) at two sites in central Canada. If a below-ground NPP allocation of 40% in both species were assumed, these are similar to the results reported here for aspen (0.73 - 2.15 g MJ^{-1}) but somewhat higher than the values for spruce (0.28 - 1.48 g MJ^{-1}). Approximations of   for total production (assuming the below-ground allocation is reasonable) demonstrate that aspen can utilize   IPAR more efficiently than Hunt and Running s model results suggest, whereas spruce show the opposite trend: can be substantially lower than their results suggest. These within-species variations in   are more than half the range of variation observed in both field measurements and model simulations across a wide range of forest stands.

Within-Species Variability in PAR Utilization

The functional convergence hypothesis (Field 1991) suggests that evolutionary optimization of resource utilization should result in similar PAR utilization efficiency among plant communities. Measurements of the relationship between net ecosystem exchange of CO_2 and solar flux in a tropical forest was found to be "virtually the same" as that observed in temperate (Baldocchi et al. 1987) and boreal (Desjardins et al. 1985) ecosystems in unstressed conditions over the very short term (on the order of hours) (Fan et al. 1990). This observation is consistent with the hypothesis of functional convergence.

Based on the results reported here and the comparison with measured values of   in other forest ecosystems, however, it was clear that   for above-ground production varied, at least between 0.07 - 1.40 gMJ^{-1}. The results of the remote sensing approach demonstrate that variations in   within species are nearly as large. Simulated and measured values of   over a range of vegetation functional types demonstrate an even larger variability. A similar result was noted within and between species in stomatal conductance (K rner et al. 1979). These results suggest there is not a functional convergence in PAR utilization efficiency. Although measured values of   are rarely representative of entire communities (which often omit below-ground production, dead and decayed matter, herbivory, and perhaps various canopy layers and ground cover), there is still evidence that   varies over a substantial range.

What might be the reasons for the observed variability, particularly between different stands of the same species? Landsberg (1986) discussed temperature, mineral nutrition and water stress factors that may modify radiation use efficiency within species. Because all the estimates of   in the study area were derived over the same five year time period, temperature effects can be ruled out. Similarly, all vegetation in the study area utilizes C3 photosynthesis, which eliminates differences due to carbon fixation pathway. Of the remaining possible causes, five likely ones were examined: (i) variation in production by other components of the stand, such as understory and ground cover vegetation; (ii) variation in carbon losses to death, decay, and herbivory; (iii) variation in below-ground allocation; (iv) variation in stresses associated with limiting resources (including mineral nutrition and water limitations); (v) variation in respiration demands. Because the first two possible causes are quickly eliminated, they are considered together.

Variation in Stand Composition and Physical Carbon Losses

Variation in carbon losses due to differences in the relative amounts of death, decay, herbivory, or understory and ground cover productivity among stands is not likely for two reasons; (i) the young aspen stands observed in this study had little to no understory or ground cover component, and little to no dead canopy matter, yet they exhibit large variations in both NPP and   , (ii) although the productivity of other stand components were not measured, other properties such as composition, density and standing dead matter did not vary between stands in any manner that would explain the observed trends in   .

Variation in Below-Ground Allocation

It is difficult to address the relative allocation of carbon to above-ground versus below-ground production in tree species due to the difficulty in making below-ground measurements. Santantonio et al. (1977) report a consistent relationship between below-ground biomass and DBH for a wide range of tree species in a variety of environmental conditions, which suggests a constant proportion between above and below-ground biomass. Widely utilized ecophysiological models generally assume a constant proportion of above and below-ground production (Bonan 1991a; Running and Hunt 1993). Furthermore, Raich and Nadelhoffer (1989) found a strong positive correlation between above and below-ground production using soil carbon budget simulations with an ecophysiological model. As Raich and Nadelhoffer point out, however, there is considerable disagreement on this subject. One of the two available data sets of total production in aspen (Ruark and Bockheim 1987, 1988) suggests that the ratio of below- to above-ground production in aspen decreases significantly with age. In the other data set, Alban et al. (1978) measured below-ground biomass of only two aspen trees of similar age, from which no generalizations can be made. As a result, it was not possible to conclude that there is a greater allocation of carbon below-ground in stands of low above-ground NPP.

Variation in Available Resources

Surveys compiled for a wide range of forest stands demonstrate that allocation tradeoffs and associated life-history traits vary with resource availability (Loehle 1988; Reich et al. 1992; Chapin 1993). Quaking aspen is adapted to relatively resource-rich environments, occupying sites where allocation to light capture is selected (Pastor and Bockheim 1984). Much of the variability in aspen NPP has been shown to be closely associated with site quality (soil depth, texture, nutrients and moisture holding capacity) (Koerper and Richardson 1980; Lieffers and Campbell 1984). In comparison, black spruce is a stress-adapted species, occupying relatively cold, wet, low-pH environments where poor nutrient uptake restricts growth (Van Cleve et al. 1983; Lieffers and MacDonald 1990). Most importantly, much of the variability in NPP among spruce stands has been shown to be related to variations in nutrient uptake that are manifested in foliage display (Van Cleve et al. 1983; Viereck et al. 1983; Pastor et al. 1987). McGuire et al. (1992) and Melillo et al. (1993) demonstrated such nitrogen limitation of production is particularly strong in boreal and tundra ecosystems.

These results suggest that within-species variations in NPP are due to greater stresses in resource-poor sites than in resource-rich sites, and more importantly, that the differences in NPP should be evident in foliage display. The positive correlation between IPAR and AANPP in spruce and young aspen stands supports this observation. It likely reflects a requirement that sufficient nutrients be available to permit a specific level of PAR interception. If this were a general explanation, resource availability would be expected to have little effect on in the study area, as predicted by the functional convergence hypothesis. Nevertheless, periodic moisture stress regulated predominantly through stomatal control rather than through diminished LAI (Verma et al. 1986; Iacobelli and McCaughey 1993) results in a similar effect. Such a strategy allows short-term moisture stress to decouple PAR interception and NPP, unlike a species that regulates predominantly through foliage construction. Periodic moisture stress may therefore mask the relationship between nutrient availability, foliage display and NPP (thus between IPAR and NPP).

Furthermore, Verma et al. (1986) have shown that forest canopies are more closely coupled with the atmosphere via turbulent mixing (higher aerodynamic roughness and boundary layer conductance) than grasslands and crops, and are thereby less closely coupled to the net radiation environment (the omega factor) (McNaughton and Jarvis 1991). From these observations, it is apparent that within-species variations in   may be driven by variations in resource (particularly water) availability. This is most evident in aspen, which are located on sites subject to periodic drought stress.

There was also an age dependence in PAR utilization efficiency of aspen stands. Mature stands in particular exhibit a lack of sensitivity in the IPAR - AANPP relationship. The factors considered to this point are insufficient to explain this observation.

Variation in Respiration Demands

Variation in the respiration demands between stands may result from increases in the relative proportion of foliar to total biomass within a stand (owing to increased costs associated with maintenance of non-photosynthetic material) (McCree 1974; Biscoe et al. 1975). In forest ecosystems, the proportion of assimilate losses through respiration processes may increase with stand age as the proportion of non-photosynthetic (woody) to photosynthetic (foliage) biomass increases (Waring 1983; Ryan and Waring 1992).

In spruce, there was no indication of variation in   related to stand age. As noted, spruce is a stress-adapted and long-lived species in which foliage display is tightly linked to stand productivity. Presumably, in the stands observed here, variations in   related to age were small relative to variations due to the other factors discussed thus far. This observation also supports the assertion that increased nutrient availability offsets increased respiratory costs (McGuire et al. 1992).

In contrast, model simulations suggest that age dependent variations in respiration are the primary determinant of variability in   in aspen stands (Hunt and Running 1992b). The results reported here (see Figure 7) are consistent with the observation that maintenance respiration losses progressively affect production rates in aspen as stands age. Nevertheless, there were mature aspen stands that had high values of   , as well as young aspen stands with low values of   . The estimates of standing above-ground biomass and bark area index were poor descriptors of this variation ($r^2=0.07$, $p<0.5$ and $r^2=0.17$, $p<0.1$ respectively). Ryan and Waring (1992) found such traditional measures of maintenance respiration (standing above-ground biomass and bole surface area) may be subject to large errors, relative to a simple model based on sapwood volume and temperature. Unfortunately, sapwood measurements were not available to explore this further, and sapwood estimated from total above-ground biomass was no better related.

The observed variability of   suggests that resource-use optimization through functional convergence (Field 1991) may not result in a narrow range of   (**Figure 8**). It is likely that a combination of drought stress and differences in respiration demands caused the observed variations in   . This conclusion was explored further with an ecophysiological model, as described in the next chapter.

Regional NPP Estimation with Constant PAR Utilization

To examine how significant the errors associated with   for estimation of NPP with remotely sensed data are, and whether a constant value of   can be used to estimate NPP over large areas, the median values of   for spruce and aspen stands were compared with independent   values generalized for "broadleaf deciduous trees" and "needleleaf evergreen trees" (Potter et al. 1993). The analysis of Potter et al. (1993) with a very narrow range of   values results in NPP estimates at a global scale that are consistent with other approaches, which is highly encouraging. The values of   estimated with their CASA (Carnegie-Ames-Stanford Approach) model for the forest classes most closely associated with the stands observed here were 0.510 g MJ^{-1} for broadleaf deciduous and 0.568 g MJ^{-1} for needleleaf evergreen. These values were derived from a doubling of their reported values to convert from g C to g dry matter (i.e., assuming a 50% carbon content in dry matter). The median values derived from remote sensing were 0.904 g MJ^{-1} for broadleaf deciduous (aspen) stands and 0.452 g MJ^{-1} for needleleaf evergreen (spruce). Entering these four different values of   into the simple model defined by Equation 1, AANPP was predicted and compared to the observed values.

Assuming the value of   Potter et al. (1993) derived for the broadleaf deciduous class results in a mean underestimation of measured AANPP in the aspen stands (ranging from -670 g m^{-2} yr^{-1} to +43 g m^{-2} yr^{-1}) (Figure 9). The value of   they derived for the needleleaf evergreen class resulted in estimates of AANPP much closer to those measured in the spruce stands, ranging from -134 g m^{-2} yr^{-1} to +288 g m^{-2} yr^{-1}. Adjusting their values to reflect above-ground production alone (assuming above-ground is 50% of total) results in a greater net underestimation (=0.255 in Figure 9). These results do not suggest that the estimates of   in either study are in error, but they do provide an indication of the magnitude of errors that can be encountered in applying estimates of   derived for one ecosystem component (i.e., trees) to an entire vegetation community, as well as errors associated with observations at one spatial scale to estimate production at another. The scale dependence of   is therefore a subject that should be addressed when more complete data sets become available.

The estimates of AANPP predicted with the median value of   observed in aspen stands ranged from an underestimation of 281 g m^{-2} yr^{-1} to an overestimation of 276 g m^{-2} yr^{-1}. In spruce, the median value of   observed resulted in a range of estimates between -197 and

+205 g m^{-2} yr^{-1} of the observed AANPP. As expected, based on the use of median values of ,
the mean residual of the AANPP estimate was small (-16 g m^{-2} yr^{-1} in aspen and -30 g m^{-2} yr^{-1}
in spruce). There were, however, substantial errors associated with the estimates for each species
(RMSE = 115 g m^{-2} yr^{-1} in aspen and 71 g m^{-2} yr^{-1} in spruce).

Using these median values of , and assuming the stands observed here are statistically
representative of the study area, a spatial extrapolation to those portions of the study area
occupied by cover types characterized as broadleaf deciduous (499 km^2) and needleleaf
evergreen (424 km^2) (based on the classification of Hall et al. 1991a) resulted in an estimate of net
annual carbon drawdown of 256 Mg C km^{-2}. This partitions into 363 Mg C km^{-2} for broadleaf
deciduous areas and 125 Mg C km^{-2} for needleleaf evergreen areas. These estimates are for the
above-ground component of the forest canopy only, and assume a 50% carbon content in the
AANPP estimates. The net contribution of other ecosystem components (mixed cover types,
peatlands, etc.) would have to be included to consider these values representative of a complete
boreal ecosystem. Nevertheless, this estimate compares reasonably well with net ecosystem
production derived from a combined ecosystem simulation model and remote sensing approach
in Alaska (Bonan 1993a).

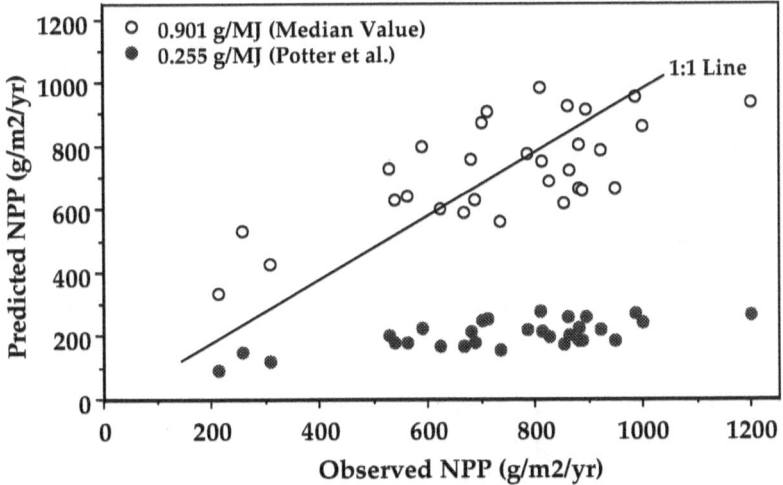

Figure 9. Observed AANPP compared to predicted AANPP derived from IPAR and $_i$
values for biomes (Potter et al. 1993) and stands (using median values $_i$ values reported
here).

Summary of Remote Sensing Analyses

Annual estimates of PAR interception derived from high spatial resolution satellite data, f_{par} and temperature-driven phenology models of boreal forest stands demonstrated both the capability and the limitations of remote sensing for estimating AANPP and PAR utilization efficiency over a range of stand age classes and site conditions. Relationships between G_i and f_{par} were species-specific and highly dependent on background and understory vegetation. Thus assumptions of canopy PAR interception proportional to SVI may result in large errors in the estimation of NPP in situations where NPP data are not available for all constituents that contribute to stand reflectance.

Errors in the NPP estimates may also result when foliage display is decoupled from annual production, which can occur through intermittent resource limitations (predominantly moisture stress) and respiration losses associated with stand age. The weakest relationship between IPAR and AANPP occurred in mature aspen stands, a result which is consistent with increased respiratory costs relative to photosynthetic gains. In contrast, strong positive relationships between IPAR and AANPP were noted in lowland spruce stands and in young aspen stands, and there was no age dependence in the relationship for spruce. Remotely sensed estimates of light interception for these stands can be used with confidence to extend NPP estimates over larger areas if canopy interception is properly estimated from SVI.

The range in PAR utilization efficiency ( ) was within the range of previously reported values for forest ecosystems, particularly temperate species. On average, values of   in spruce stands were about half those of aspen. Moreover, the range of   within species was nearly as large as between species, particularly among young aspen stands. This result may reflect successional replacement of unproductive stands as they age. Reasons for observed variations in   are explored further in following chapters. The estimates of   include errors propagated from various sources, including both measurement and modeling errors (see results section), however, the fact that the range of values were within those of comparable stands with similar radiation measurements suggest the results are robust and independently derived estimates of   reported in the next chapter support this conclusion.

These results suggest that vegetation functional type differences (e.g., broadleaf deciduous versus needleleaf evergreen) should be considered in the remote sensing of NPP, not only in the development of f_{par} models, but also in quantifying radiation use efficiency. Physiological differences between functional types has also been noted as a primary determinant of net ecosystem production in ecophysiological modeling approaches (Bonan 1993a).

Areal estimates of net carbon uptake over the 2280 km^2 study area, derived from the IPAR data and median values of   , compare favorably with estimates of net ecosystem production derived for another boreal forest ecosystem (Bonan 1993a). Estimates of AANPP based on representative values of   , derived using the CASA global-scale model underestimated observed production at the stand-level, particularly in aspen. In so far as the CASA model is reliable, this was attributed to differences in the spatial scale of the observations (biome-level versus stand-level), and differences in whole ecosystem versus single component (tree) estimates of AANPP and   .

Within-species variations in   of the magnitude reported here suggest that optimization of resource use through "functional convergence" may not result in a convergence on a very narrow range of values. A literature survey of   values and ecophysiological model simulations of   in a variety of ecosystems support this conclusion (Figure 8). The results from this analysis suggest that accounting for stresses and respiration components will be required in NPP models based on a remote sensing (i.e., light harvesting) approaches, as they were originally proposed (Monteith 1972; 1977) and widely utilized.

As a consequence of these conclusions, sources of variability in   were explored in a more general context using a mechanistic ecophysiological model, as reported in the next chapter.

Chapter III. Modeling Carbon Fluxes in Boreal Forest Stands

In this chaper a mechanistic ecophysiological (carbon flux) model is used to more thoroughly examine the observed variability in   . The Terrestrial Carbon Exchange (TCX) model was selected for these analyses because it was designed for operation at the stand level, could be parameterized with the stand data available for this study, and most importantly, because it was developed for and has been extensively tested in boreal forest ecosystems (see Chapter I).

Terrestrial Carbon Exchange (TCX) Model Description

The TCX model was developed for stand-level studies in boreal forest ecosystems using linked carbon (Bonan 1991a; 1993b) and surface energy budget (Bonan 1991b) routines. The subroutine calling sequence and their respective products are listed in Table 5. There are four basic components of the model: one that describes surface energy budget (SEB) and water fluxes, and one each to describe the carbon fluxes for trees, moss and soil. These are briefly described here, but the reader is referred to the publications describing the model for specifics (i.e., Bonan 1991a, 1991b, 1993b). The SEB component of the model calculates energy, heat, momentum and moisture fluxes between forests and the atmosphere, and heat and moisture fluxes in the soil. The SEB terms are linked to calculations of decomposition, nitrogen (N) mineralization, carbon exchange (photosynthesis and autotrophic respiration) for trees and moss, and heterotrophic respiration for soils (R_h).

Input and output variables are listed in Table 6. Climate data are used to drive the model simulations on a daily basis, subdivided into hourly intervals. A period of three months of climate data (typically October through December) is used to "spin up" arbitrarily specified initial conditions of soil moisture and temperature. The model is static in the sense that LAI, sapwood biomass, root biomass and soil carbon parameters are all specified, and it runs only on an annual basis. That is, multiple-year runs require model reinitialization for each annual simulation period.

Irradiance incident at the surface is calculated from top-of-atmosphere irradiance reduced by fractional cloud cover. A two-stream model is used to calculate canopy absorption, reflectance and transmittance of light, modified by canopy spectral properties, for three-layers (upper and lower canopy, and ground surface). PAR absorption in each canopy layer is based on exponential light extinction through the layers and reflection from the ground surface.

Soil moisture and temperature profiles are also calculated for a series of layers, and foliage temperature, water potential and vapor pressure deficit, are updated on an hourly basis. These variables are used, together with absorbed PAR, to calculate stomatal conductances by

48

employing a threshold response to the most limiting factor and assuming they operate independently (after Jarvis 1976, Ball *et al.* 1987). Net radiation, sensible and latent heat fluxes are solved simultaneously at each height and updated for the multi-layer canopy and soil. If present, snow, moss and humus are also updated.

Year Loop:
Specify constants (CONSTS)
Initialize parameters (INIT)
Read in control data (CONTROL)
Read in climate data (CLIMATE)

Day Loop:
Calculate solar radiation on surface (SUN)
Simulate deciduous phenology (LEAFOUT)
Characterize soil root abundance (ROOTS, SAF)
Calculate foliage water potential (FWP)
Evaluate surface energy balance (SEB)
   Emissivity of air (EMISS)
   Net radiation transfer coeffs (COEFFRN)
   Soil surface albedo (SURFALB)
   (Call Layer Loop)
Calculate canopy and surface water storage (CWS)
Update snow pack (MELTSNO)
Update soil temperature (SOILT)
Update soil water (SOILW)
Calculate carbon exchange (CARBON)
Output results (OUTPUT1-4)

Layer Loop:
Incident PAR and incident and absorbed total (SOLAR)
Two-stream reflectance (TWOSTR)
Wind speed (WIND)
Sensible heat resistances (RESSEN)
Surface resistance (SURFRES)
Sensible heat coefficients (COEFFSH)
Canopy air temperature (CANATMP)
Stomatal resistance (STOMATA)
Water vapor resistances (RESWVP)
Water vapor coefficients (COEFFWV)
Canopy air pressure and evaporation (CANAWVP)
Calculate net radiation (NETRAD)
Calculate sensible heat flux (SENHEAT)
Calculate transpiration (WATRVAP)
Calculate soil heat flux (SOILH)
Calculate partial derivatives (DERIVS)
Update temperatures (SOLVE)
(iterative solution on CANATMP)
Calculate relative humidity (RELHUM)

Table 5. Sequence of routines called in TCX model. Subroutine name is in parentheses.

49

```
                              Input  Data

        Biotic Variables                          Physical Variables
        Vegetation type                      Soil color (light to dark)
       (deciduous or conifer)              Mineral soil type (coarse, fine)
       # Vegetation Layers                        Drainage class
         Canopy Height                               Slope
           Foliage N                                Elevation
         Forest floor N                              Aspect
          Total LAI
         Depth of humus                     Meteorological Variables:
      Presence of moss (y/n)                      Air pressure
        Depth of live moss                       Air temperature
       Green moss biomass                     Dew point temperature
        Sapwood biomass                           Wind speed
       Forest floor biomass                    Fractional cloudiness
          Root biomass                            Precipitation


                              Output  Data
```

Annual C-Uptake (g dry matter m^{-2} yr^{-1}) Daily Energy Budget (W m^{-2})
  Above-ground tree production   Albedo (unitless)
       Root production   Incident solar radiation
   Moss production (if applicable)   Net radiation
    Forest floor decomposition   Evaporation flux
                                             Transpiration flux
Annual C-Release (g CO$_2$ m^{-2} yr^{-1})   Sensible heat flux

      Foliar respiration   Soil heat flux
     Sapwood respiration
                                      Stomatal conductance (m s^{-1})
       Root respiration
       Moss respiration
     Microbial respiration   Net Ecosystem C-Flux (g CO$_2$ m^{-2} yr^{-1})

Table 6. Input variables required and output variables simulated by the TCX model. Italics indicate those input variables for which measurements were unavailable, hence were estimated.

Foliage phenology is driven by soil degree days (SDD), in much the same way as ambient air temperature was used to derive growing degree days in the remote sensing analyses (see Chapter II, page 25). Photosynthesis can occur if soil temperature increases above freezing, but in deciduous stands also requires that snow cover be absent .

Photosynthesis is modeled as a function of enzyme kinetics and CO$_2$ diffusion (after Farquhar 1989). Biochemical reactions and electron transport rates are constrained by foliage temperature, N availability and PAR, which are mediated through mesophyll resistance (after Running and Couglan 1988). RuBP carboxylase (Rubisco) acts to catalyze the carboxylation and oxygenation of RuBP and the regeneration of RuBP via electron transport (i.e., C3 photosynthesis). Assimilation is limited by Rubisco availability, RuBP regeneration rates of carboxylation, CO_2 compensation point and the partial pressure of CO_2 in the chloroplast. Rubisco availability is determined by nitrogen uptake. Intercellular CO_2 concentration is determined by ambient CO_2 partial pressure, air pressure, and the resistance to water vapor diffusion (both boundary layer and stomatal). Maximum rates of carboxylation were determined from laboratory measurements and are species specific for trees.

Respiration is partitioned into foliar, stem and root components associated with both growth and maintenance of tissue. Maintenance respiration is modeled as an exponential function of temperature and varies with the mass and biochemical composition of plant tissues. Growth respiration is specified as a proportion of dry matter production and varies with the biochemical composition of the plant tissues (i.e., foliage, stem and root). Root maintenance respiration is modified by soil temperature, root biomass, and relative root abundance distributed over a series of soil layers (three layers were used). Microbial respiration is determined by soil temperature, available moisture and litter quality (C:N ratio).

The model recognizes linkages among physiology, litter quality and N mineralization and growth is constrained so that a critical C:N ratio is maintained. Available N is the sum of N fixation, N contained in precipitation and N mineralization. N content declines with foliage and stem age, and with root size, and is accumulated and stored in tissue (not lost through runoff or subsurface flow). N is used from storage before additional uptake.

Decomposition and N mineralization are determined by litter quality, which declines through time and by species, and microbial growth rates are affected by soil moisture and temperature. Litter N initially increases with time via microbial mobilization, and then decreases due to net mineralization. Different carbon pools are recognized (foliage, stem and root litter, forest floor, and mineral soil organic C). Fifty percent of foliage N is absorbed prior to abcission, but there is no resorption of root litter N.

Finally, net photosynthesis is obtained by differencing C assimilation and respiration, and NPP is calculated as the annual sum of net photosynthesis reduced by a factor of 0.44 to account for the fractional molecular mass of C in CO_2.

Methodological Approach

In order to use of the TCX model to analyze variability in   , the following steps were necessary: (i) parameterize the TCX model with measured stand population characteristics, (ii) operate the model to converge on observed NPP values at the sites, (iii) analyze the variability in modeled carbon fluxes, R:A ratio, and   among stands. Before these steps could be taken the model was first modified to allow input of several measured variables that were available, and then it was subject to a series of sensitivity tests. The approach is outlined in Figure 10.

In order to use as much measured data as was available for the sites, instead of simulating all the input variables, TCX was modified to ingest atmospheric relative humidity and incident irradiance on a daily basis. The model was also modified to specify species-specific stem and root respiration coefficients (after Hunt and Running 1992b) and to calculate air pressure instead of requiring it as an input variable. Information on soil properties was modified based on field measurements (after Levine et al. 1993). Each of these modifications is described below.

Because measured values of incident irradiance were not partitioned into direct and diffuse components, the use of fractional cloud cover, and atmospheric transmission and emissivity calculations based on it, were retained. Instead of being specified, however, fractional cloud cover was retrieved using measured surface irradiance and simulated top-of-atmosphere irradiance (at the site latitude and longitude) to calculate atmospheric transmission ( ). Fractional cloud cover was then empirically related to   by inverting the model transmission calculation. The model was exercised for a range of fractional cloud cover values to determine the relationship to direct and diffuse PAR irradiance (Figure 11). Thus, although measured values of total incident PAR were used in model calculations, the partitioning of irradiance into direct and diffuse components was done using the original model structure.

To avoid the requirement for input air pressure (P_a) measurements (which were unavailable) P_a was calculated on a daily basis by adjusting the pressure of a standard atmosphere with the ratio of observed average daily air temperature and standard atmosphere temperature (i.e., Charles Law). Air pressure is used in the model for minor adjustments of atmospheric emissivity calculations, which affect the partitioning of incident PAR into direct and diffuse components, as described above.

The values of soil bulk density and volumetric water content were specified based on field samples from the spruce and aspen stands, with the assistance of soil scientist Dr. Elissa Levine. The bulk density of fine mineral soils (i.e., aspen stands) was increased from pre-specified values, and the volumetric water content of peat (i.e., spruce stands) was added to the soil properties specied in the model.

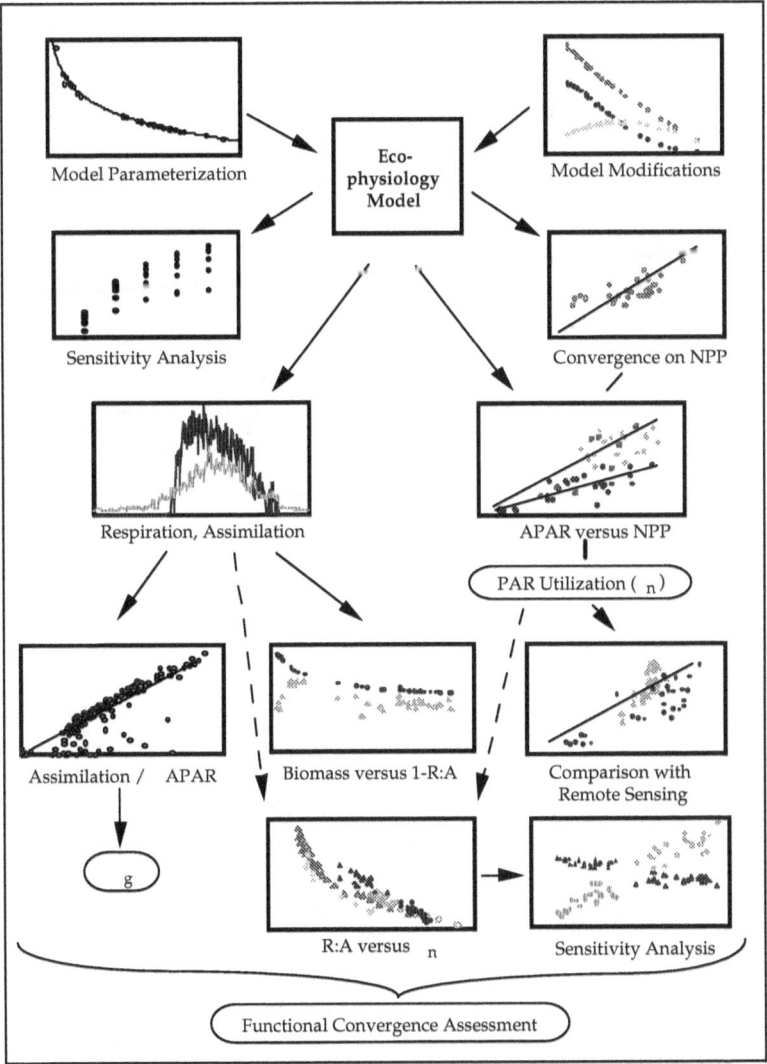

Figure 10. Flowchart of the methodological approach used for the carbon flux modeling portion of the analysis.

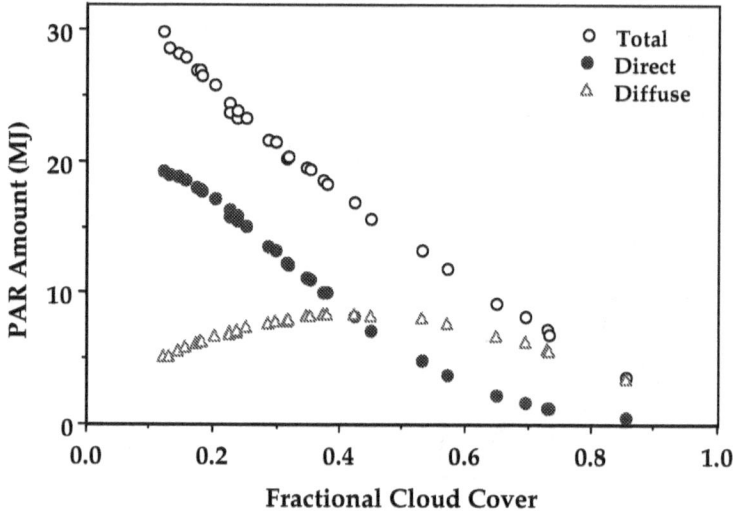

Figure 11. Diffuse, direct and total daily PAR amounts (MJ) as a function of fractional cloud cover.

Sensitivity Analysis

A sensitivity analysis of the model had previously been conducted using a Monte Carlo simulation approach (Bonan 1991a; Chapter I), however, owing to model modifications and different objectives, an additional sensitivity study was conducted for the research reported here. Moreover, it was necessary to characterize the sensitivity of the model to variables for which measured input data were not available (indicated in Table 6). The sensitivity analysis was conducted by varying the value of input variables, over a reasonable range, in a stepwise fashion. The sensitivity of the model to the input variables provided an indication of their relative importance to NPP evaluation. There were no formal statistical analyses of the results as their purpose was primarily to assess the performance of the model and to characterize the importance of variables for which measurements were unavailable. The sensitivity analyses were , however, also used to examine variability in , as discussed later, and these were tested for significance.

Parameterization

The TCX model was parameterized on a site-specific basis using measured or calculated stand characteristics (vegetation type, canopy height, canopy LAI, sapwood biomass, root biomass, etc.) and site properties (elevation, aspect, slope, soil properties) (Table 6). When measured variables were unavailable, literature values for boreal forest stands in Alaska (Bonan 1991a; 1993b) were used. Some measured variables (such as above-ground biomass and age) were used to calculate variables required by the model (such as root and sapwood biomass amounts), as described below.

Maintenance respiration coefficients were specified uniquely for foliage, stem and roots based on laboratory measurements (Bonan 1991a). An additional parameter was added to distinguish between coarse and fine roots (after Hunt and Running 1992b) (Table 7). Growth respiration coefficients were initially specified by Bonan (1991a) at 35%, 30% and 35% for foliage, stem and root, respectively. Bonan (1993b) later found these were over-estimated and adjusted them each to 25%. Hunt and Running (1992b) used values of 30%, 25% and 30% for foliage, stem and roots, respectively. Bonan s 1993 values were used based on his success with the model in reproducing field measurements. Thus growth respiration coefficients were specified as a constant 25% of assimilation for each of foliage, stem, and roots, for both species. While the growth respiration coefficients can greatly alter the NPP results, they affect all sites in the same manner and thus do not affect the comparative analyses. Carbon allocation to foliage, stem and roots was specified by species (Table 7).

| | Respiration | Parameters | | | Carbon | Allocation |
| | Spruce | Spruce | Aspen | Aspen | Spruce | Aspen |
	$\frac{mg\ CO_2}{kg\ s}$	Q_{10}	$\frac{mg\ CO_2}{kg\ s}$	Q_{10}	%	%
Foliage	0.3584	1.78	0.0937	1.86	20	37
Above-ground Woody	0.0428	2.34	0.0428	2.34	40	38
Fine Roots	0.1650	2.51	0.1650	2.51	20	12.5
Coarse Roots	0.0823	2.51	0.0823	2.51	20	12.5

Table 7. Maintenance respiration coefficients and carbon allocation parameters specified for the carbon flux model simulations.

Below-ground biomass in aspen was estimated from measured total above-ground biomass and the relationship with stand age derived from data reported for an age sequence of *Populus tremuloides* in northern Minnesota (Ruark and Bockheim 1987, 1988). The resulting allocation of biomass below-ground ranged from 20% - 50% of above-ground biomass (Figure 12a), which is consistent with values reported by (Santantonio et al. 1977) for temperate deciduous species. Bonan (1993b) and Hunt and Running (1992b) estimated below-ground biomass in spruce at a constant 66% of above-ground. While all the spruce stands in the SNF study area were lowland sites, some were clearly better drained than others and had correspondingly higher AANPP. Below-ground biomass in spruce was thus specified between 50-90% of above-ground depending on the relative soil wetness of the site. The derived values were consistent with those reported by Santantonio et al. (1977) for temperate conifer species.

As with below-ground allocation, sapwood biomass in aspen was estimated from measured total above-ground biomass and its relationship with stand age as reported by Ruark and Bockheim (1988). Using this approach allowed a realistic approximation of sapwood:above-ground ratio relative to stand age. Estimated values ranged from 17% - 77% (Figure 12b), which is consistent with other age-specific observations (Ryan et al. 1994) and with measurements collected as part of the Boreal Ecosystem Atmosphere Study (BOREAS) in central Canada (at ~15% for ~80 year-old aspen stands; J. Vogel, personal communication). Sapwood biomass in spruce stands was estimated at a constant 12% of above-ground biomass based on BOREAS measurements at comparable sites. Bonan (1993b) and Hunt and Running (1992b) estimated it at a constant 26%. Estimating sapwood as constant proportion of above-ground biomass in lowland black spruce is not an unreasonable assumption owing to the life history traits of the species (i.e., slow growing, stress tolerant) (Ryan 1990).

Foliage biomass was not a required input variable for the model because it is calculated from LAI (at 286 g foliage biomass per m^2 LAI, after Landsberg 1986). For comparison with measurements made at the BOREAS sites and values calculated as a proportion of LAI, however, foliage biomass was calculated for aspen using the age-specific measurements of Ruark and Bockheim (1988). The derived values ranged from 1.0% - 7.1% of total above-ground biomass, compared to 2.2% - 6.1% in the data collected by Ruark and Bockheim. In both cases foliage biomass production was a nearly constant 25% of AANPP. The foliage biomass amounts were consistent with the BOREAS measurements, which ranged from 1.0% - 1.9% in mature stands (J. Vogel, personal communication). In contrast, estimates of aspen foliage biomass calculated as 286 g m^{-2} LAI exceeded total above-ground biomass in many cases, particularly young stands. Based on the favorable comparison of foliage biomass amount when calculated as 25% of above-ground AANPP for the aspen stands, this method was used to derive a mean values of 65 g

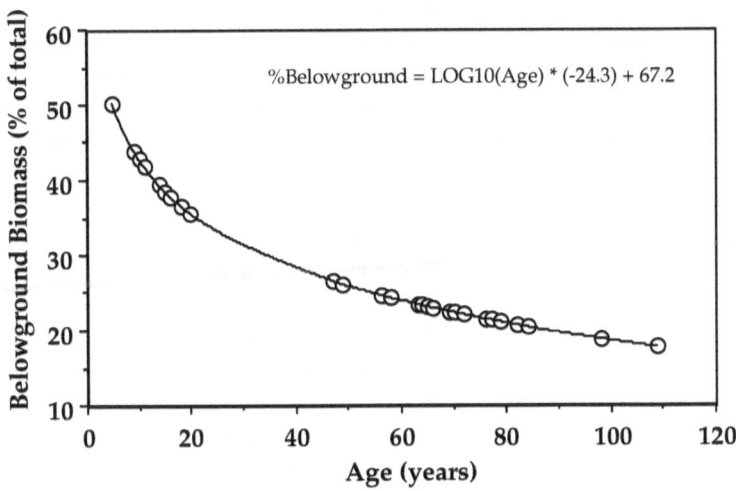

Figure 12a. Age dependence of aspen below-ground biomass amount as derived from the analysis of Ruark and Bockheim (1988). Circles indicate the individual stands used for this work.

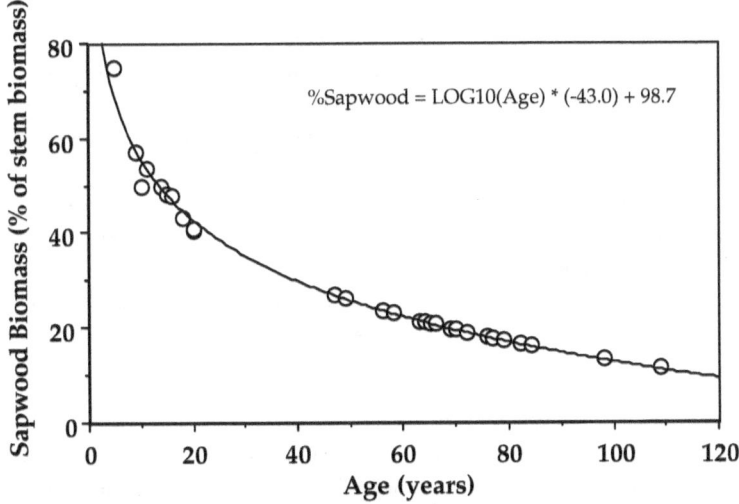

Figure 12b. Same as Figure 12a but for aspen sapwood biomass amount.

biomass per m^2 LAI, which was reasonably consistent among stands (= ± 15 g m^{-2}, N=30). Estimating foliage biomass as 50% of above-ground (equivalent to 37% of total production assuming 25% below-ground allocation; Table 7) resulted in an estimate of 95 g m^{-2} LAI. Thus, the model was modified to calculate foliage biomass in aspen stands as the latter of these two figures, i.e., 95 g m^{-2} LAI, to be consistent with the foliage allocation specified by both Bonan (1993b) and Hunt and Running (1992b) (i.e., 37% of total production). The effect of this modification is consistent among stands and does not alter between-stand comparisons. Estimating a comparable value of foliage biomass in relation to the NPP of spruce stands is complicated by the retention of foliage from year to year (i.e., annual foliage NPP is more difficult to measure). Thus the original figure used by Bonan (1993b) (286 g m^{-2} LAI) was retained for the spruce stands. The higher leaf mass per unit area in spruce is consistent with their evergreen habit and structural differences associated with greater longevity.

Foliage samples were collected in the summer of 1992 but they unfortunately spoiled before funds could be procured for nutrient analyses. Thus, top-of-canopy leaf nitrogen content values were not available for the study site. Numerous literature values were available, however, and these were used to provide representative values for each species. The values were augmented with data collected as part of BOREAS (Dang et al., in press). Hunt and Running (1992b) and Bonan (1993b) used constant values of 2.1% for aspen and 0.9% for spruce. Values measured in BOREAS averaged 2.47% for aspen and 0.68% for spruce (for a sample of trees in three mature stands). Ryan (1995) reported values for aspen of 1.9% (= ± 0.72) in Colorado (n=6) and 2.3% (= ± 1.12) at the BOREAS sites (n=4). For the model simulations, the range of values among stands of the same species was allowed to vary as a free variable in order to reproduce observed AANPP. The values used in the modeling exercises were within the range with those observed in the literature and in BOREAS, as shown later.

Humus and moss depth were estimated at the sites from field observations. Humus and moss biomass and N were not known, but were approximated by the depth estimates. Humus N concentrations were estimated from foliage N, less resorption, and moss N was specified at a constant 0.83% (after Bonan 1991a). These "forest floor" variables were not crucial to the objectives of estimating tree NPP, however, they have an effect on microbial respiration rates and associated N mineralization.

Stand Simulations

The model was run to converge on observed NPP using all the variables for which stand measurements were available and modifying the only remaining parameter capable of reproducing the measured NPP values; foliage N. Simulated values of maintenance respiration and assimilation (less photorespiration) incorporated intrinsic limitations on photosynthetic rates owing to the effects of temperature and stomatal control . For example, aspen stands on thin, coarse textured mineral soils had less available soil moisture than stands on deep, fine textured soils, thus greater stomatal control was exerted which in turn reduced assimilation. The ratio of maintenance respiration and assimilation was analyzed with respect to biomass, NPP and PAR utilization. Simulated canopy APAR and PAR utilization were also compared with the production efficiency model results from Chapter II.

In addition to the model simulations run with the stand-specific parameterizations (i.e., the measured and derived stand variables), an additional set of simulations were run using a generalized range of stand variables (i.e., with all variables except LAI and foliage N held constant). For example, aspen stand LAI was varied between 1 and 5, and foliage N between 1.25 and 2.5%. This allowed an analysis of the explicit (i.e., measured) stand parameterizations compared to a range of more generalized simulations.

Simulation Results

The results of the model simulations included those from the sensitivity analysis, the stand-specific parameterizations, and the generalized range of stand variables. The various results were analyzed with respect to simulated variability in the amount of net production per unit APAR ($_n$), the amount of carbon assimilation per unit APAR ($_g$), and the ratio of respiration to assimilation (i.e., "respiration efficiency").

Sensitivity Analysis

Results of the sensitivity analysis are summarized in Table 8. The sensitivity of all variables were not tested because they were not directly relevant to the objectives of this research. For example, while atmospheric CO_2 concentration had a moderately strong effect on NPP, it was not explored further. Sensitivity analysis results were ranked as weak, moderate and strong; "weak" represents to a change of <10% in simulated total tree NPP from the baseline case, "moderate" a change of 10-40%, and "strong" >40%.

Total tree NPP was weakly dependent on variables such as soil color and moderately dependent on variables such as sapwood biomass, soil type (e.g., properties including texture, bulk density and water holding capacity), and forest floor N content. Biomass of the forest floor mostly affected heterotrophic respiration and had little effect on tree NPP. Moss biomass affected only moss production. Tree above : below-ground biomass ratio had a moderately strong effect on NPP because it modified respiration costs. For example, increased total tree respiration resulted from increased below-ground allocation, hence in decreased NPP, because photosynthesis remained constant.

Those variables that resulted in high variability in NPP for both aspen and spruce included LAI, foliage N, and irradiance (which reflected the combined effects of variables such as fractional cloud cover and topography). Some variables affected one species more strongly than the other. For example, humus depth had only a weak effect on aspen NPP, but moss depth had a moderate effect on spruce NPP by thermally insulating the soil and so reducing the length of the growing season.

Variable	Sensitivity	
	Spruce	Aspen
LAI	strong	strong
Foliage N	strong	strong
Irradiance	strong	strong
Moss Depth	moderate	n/a
Sapwood Biomass	moderate	moderate
Root:Shoot Ratio	moderate	moderate
Soil Type	moderate	moderate
Soil Color	weak	weak
Forest Floor N	weak	weak
Humus Depth	n/a	weak

Table 8. Results of TCX sensitivity analysis with respect to total tree NPP. Response categories are described in the text (n/a indicates not applicable).

LAI had the largest effect on NPP but it elicited a different response in spruce and aspen stands. In spruce, the effect of LAI on NPP was curvilinear (Figure 13a) whereas the effect on NPP in aspen was nearly linear (Figure 13b). Spruce production was highly dependent on soil temperature, which took longer to increase to levels that permitted photosynthesis when canopy LAI was high. This effect, when combined with increased foliage respiration relative to photosynthesis, resulted in the asymptotic behavior of spruce NPP with LAI. In aspen stands the mineral soil warmed at nearly the same rate under low or high LAI, and both photosynthesis and respiration increased with LAI at a constant rate. Thus the relationship between NPP and LAI was more nearly linear.

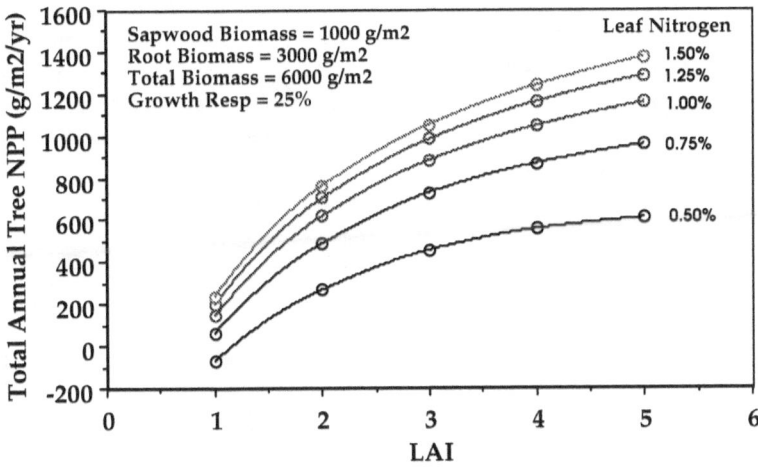

Figure 13a. Variation in simulated spruce total tree production as a result of combined variations in LAI and foliage N content. All other variables were held constant, including those indicated.

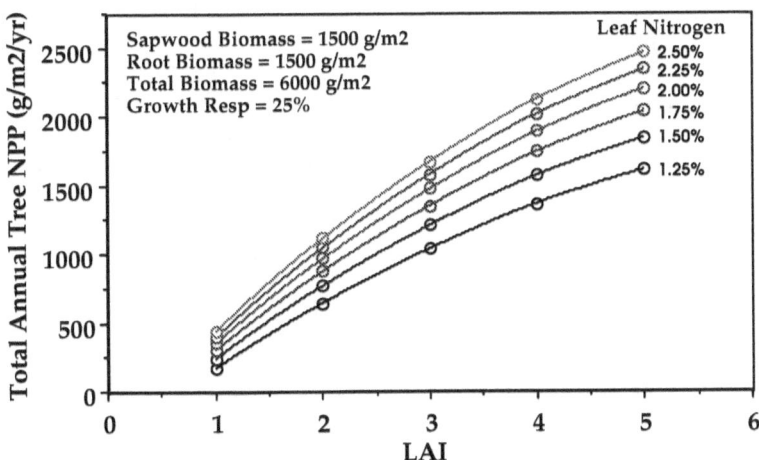

Figure 13b. Same as Figure 13a, but for aspen stands.

Foliage N increased NPP for both spruce and aspen at constant LAI (Figure 14) but was nonlinear. Spruce NPP varied more than aspen with foliage N, particularly at the lower N values (0.5 - 1.0%) (see also Figures 13a,b). This result is consistent with field observations that spruce are typically a more nutrient limited species than aspen. Note that the boundary conditions shown in Figures 13 and 14 may be wider than those that occur in nature, as analysis of the site data (next section) showed. Moreover, in the range of N concentration that occured in both species, NPP was consistently lower in spruce than aspen.

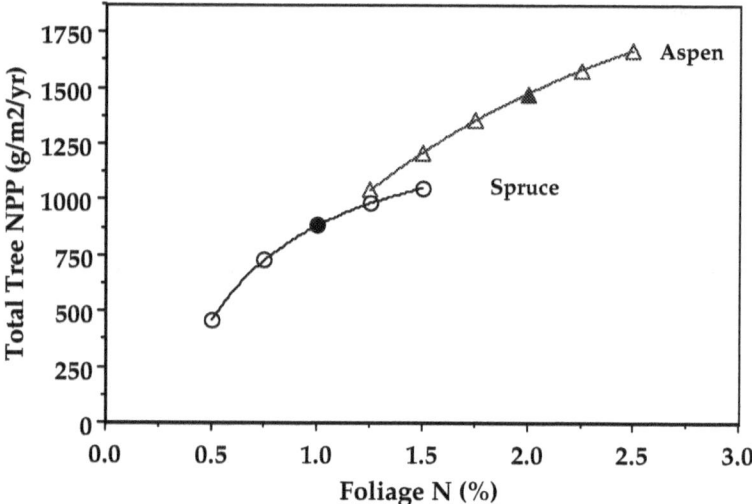

Figure 14. Variation in simulated total tree net production, at an LAI of 3.0, as a result of variations in foliage N content. All other variables were held constant, as in Figure 13. Filled symbols indicate average values of foliage N reported for the two species.

Comparison of Observed and Simulated NPP

A comparison of simulated stand NPP with observed values of NPP is shown in Figure 15a (for spruce) and Figure 15b (for aspen). The result of specifying a constant foliage N value is shown together with a 1:1 fit line that represents the solution of predicted and observed AANPP when N is tuned. A strong positive relationship between simulated and observed AANPP emerged even with constant foliage N, presumably as a result of realistic parameterization of the sites ($r^2 = 0.80$ for spruce, 0.92 for young aspen, 0.38 for mature aspen; p<0.001). The greater variability in mature aspen stands is a result of the reduced range in AANPP for these stands, as discussed earlier (Chapter II). Also, while the young aspen stands had a high coefficient of determination, most of the simulated values for low AANPP stands were overestimated. This may be a result of the smaller below-ground biomass allocation with which the young sites were parameterized, but it is more likely to be the result of inadequate characterization of moisture availability in the shallow, coarse-textured soils on which some of the low-AANPP young aspen sites were located. As noted in the last chapter, these sites incur high establishment costs and may not reach maturity owing to competitive replacement by other more moisture stress-tolerant species (typically *Pinus banksiana*).

Varying foliage N removed the remaining variability in the relationship between simulated and observed AANPP by forcing a 1:1 relationship, which was necessary before analysis of variability in $_n$ could proceed. The values of foliage N resulting from iterative solution on observed NPP were 0.4 - 1.2% (mean = 0.83%) for spruce stands and 0.7 - 2.8% (mean = 1.95%) for aspen stands. The derived foliage N values were within the range of values reported for a variety of tree species (e.g., Vitousek 1982; Field and Mooney 1986; Running and Hunt 1993). Moreover, the mean values were very close to those used by Bonan (1993) and Hunt and Running (1992) to represent typical spruce and aspen stands, at 0.9% and 2.0% respectively, and the means and ranges were very similar to those measured at BOREAS and reported by Ryan (1995), as noted earlier. Thus, they are likely to be reasonable approximations of the true values.

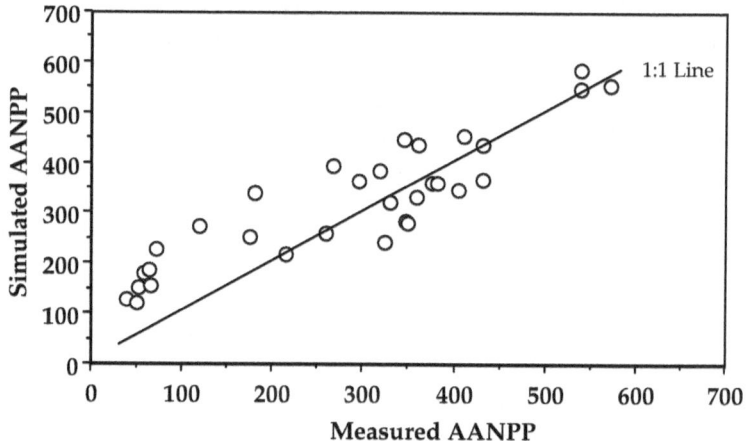

Figure 15a. Simulated AANPP for spruce stands, as determined with a constant, representative foliage N content (1.0%). The 1:1 line illustrates convergence of simulated and measured values of AANPP.

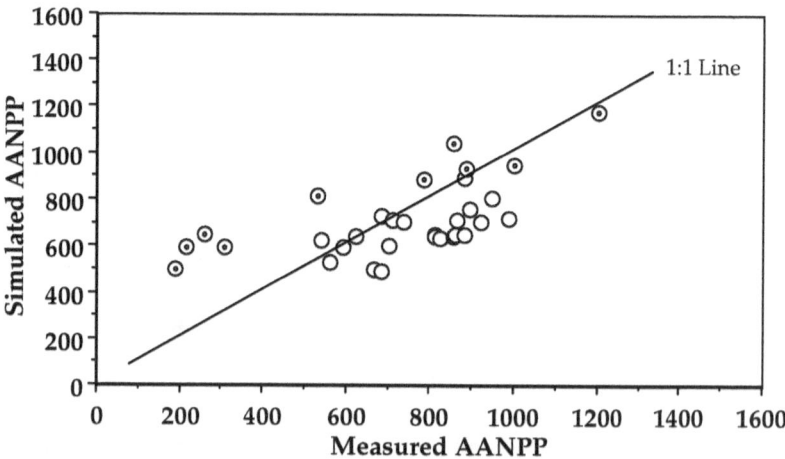

Figure 15b. Same as Figure 15a, but for aspen stands and constant N=1.9%. Young stands are indicated with double circles.

Seasonal Patterns of Assimilation and Respiration

Variability in simulated assimilation and respiration on a daily basis was large, and followed a clear pattern through the growing season in both spruce (Figure 16a) and aspen (Figure 16b). The data in Figure 16 were averaged for all the stands of each species, and are shown for a climatically typical year for the study area (1983). Note that all the analyses were conducted for a five-year period and averaged, as described in the last chapter, to match the NPP measurement period.

Maximum assimilation rates through the growing season were comparable between spruce and aspen stands, as were maximum respiration rates, but in the spruce stands a greater proportion of assimilated carbon was lost through respiratory processes. Spruce stands also showed greater respiratory losses through non-growth periods, a result of maintaining foliage throughout the year. This observation is consistent with related suites of traits among the two species in that spruce tends to allocate a greater proportion of assimilated carbon to non-growth processes, hense tends to have lower daily net photosynthesis and a slower growth rate than aspen.

The length of the growing season for this example was somewhat shorter for the spruce stands (day 132 to 315) than the aspen stands (day 116 to 313), which was a result of frozen organic soils at the spruce sites. The average number of soil degree-days was 840 for spruce and 1420 for aspen. Average soil degree-days was higher on the young aspen stands (1670) than the mature stands (1308), which is consistent with the shallower, coarse-textured soils at most of the young stands.

The simulated end of the growing season (i.e., the last day of CO_2 uptake, day ~315) was the same for the two species. This does not seem to be realistic because the aspen stands typically lost all foliage by late October (day ~300) in the study area, as the phenological observations in Chapter II showed. This may indicate an insensitivity in the simulation model to factors that determine leaf drop in aspen stands (i.e., soil degree-days and presence of snow). This result had little effect on further analyses, however, as the difference in stand averaged AANPP resulting from the extended growing season in the aspen simulations was just 14 g m^{-2} yr^{-1}, which is less than 2% of the average AANPP. Moreover, although the model was not designed to simulate it, bark photosynthesis in aspen stands could possibly account for the additional 2% AANPP at the end of the growing season.

In summary, the growing-season assimilation, respiration, net photosynthesis and soil degree-days, as simulated by the model, were reasonably realistic.

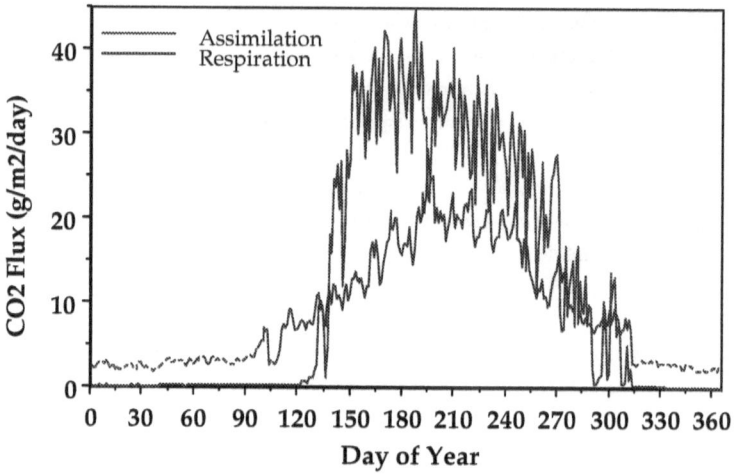

Figure 16a. Seasonal dynamics of daily assimilation and maintenance respiration, averaged for all spruce stands, for year 1983.

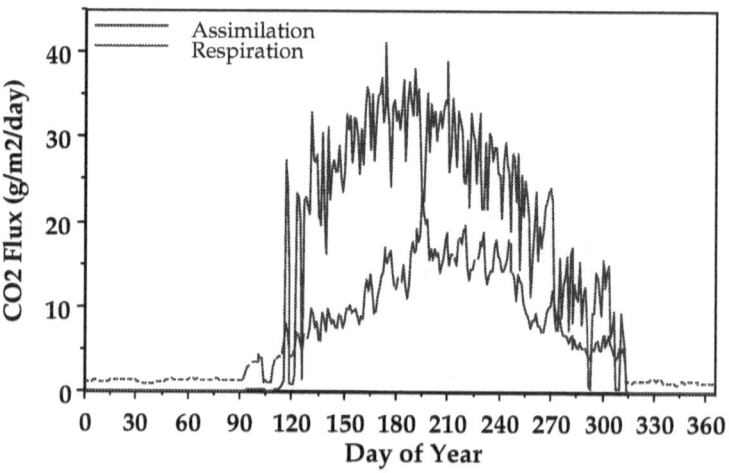

Figure 16b. Same as Figure 16a, but for aspen stands.

Sources of Variability in PAR Utilization

The relationship between simulated annual APAR and AANPP (Figure 17) was comparable to that observed with the production efficiency model (PEM) analysis (see Chapter II). In particular, the distinction between spruce and aspen stands, and the separation of young and mature aspen stands, was evident in both the model simulations and the remote sensing analyses despite their independent bases. The APAR and η results for each stand are listed in Table 9a (spruce) and Table 9b (aspen). Spruce stands averaged 0.33 ($= \pm 0.11$) g above-ground production per MJ^{-1} APAR, which was somewhat lower than the average derived from remote sensing (0.49 ± 0.17 g MJ^{-1}). Aspen stands averaged 0.91 (± 0.26) g MJ^{-1}, which was nearly identical to the value derived with remote sensing (0.92 ± 0.22 g MJ^{-1}). The range in both species ($0.15 - 0.52$ g MJ^{-1} among spruce stands and $0.35 - 1.44$ g MJ^{-1} among aspen stands) was also comparable to that observed in Chapter II (see Figure 7).

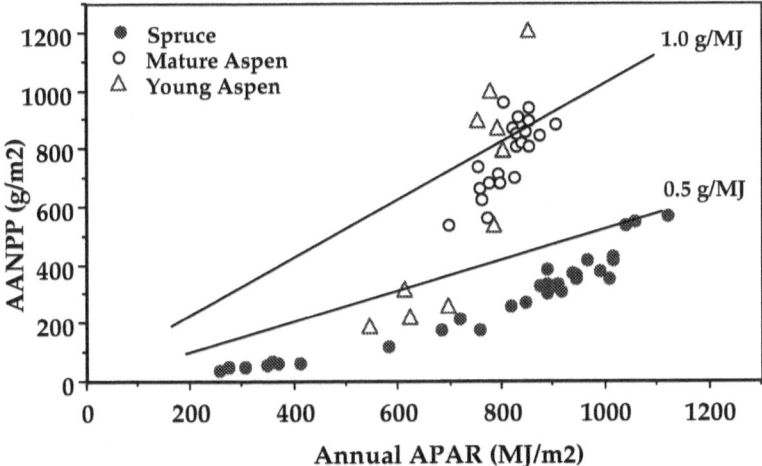

Figure 17. Simulated annually absorbed PAR and annual above-ground NPP for aspen (young and mature) and spruce stands, from simulations with measured stand variables. As in Figure 7, lines indicate dry matter yield of intercepted PAR at 0.5 and 1.0 g MJ^{-1}.

As with the production efficiency model results, APAR was a better predictor of spruce AANPP (r^2=0.90, p<0.001) than aspen AANPP (r^2=0.66, p<0.001), and the difference between spruce and aspen was significant (t=11.5, two tailed p<0.001). APAR was also a better predictor of AANPP in young aspen stands (r^2=0.76, p<0.001) than in mature stands (r^2=0.60, p<0.001), although this distinction between age classes was less clear than in the remote sensing analysis (t=1.25, p=0.23). Moreover, the low AANPP of four young aspen stands was clearly distinct from the other aspen stands, both young and mature. A regression fit of the young aspen stands, including the group of "outlier" stands, results in an unrealistic amount of annual PAR absorption (575 MJ) at the Y-intercept of zero production. These stands are discussed more later. The difference in slopes between the PEM and ecophysiological modeling approaches was driven entirely by differences in APAR calculated with the two-stream equations (in the ecophysiological model) versus IPAR calculated with the geometric-optical approach (in the PEM analysis).

The simulations of APAR spanned a comparable range to that of IPAR derived with the geometric-optical model (Figure 18). As with IPAR, APAR varied over a much greater range in spruce (256 - 1120 MJ m^{-2} yr^{-1}; Table 9a) than in aspen stands (545 - 906 MJ m^{-2} yr^{-1}; Table 9b). The IPAR values derived for the spruce stands with the geometric-optical model were, however, systematically lower than the comparable APAR values calculated from the two-stream equations. Increased PAR absorption of reflected PAR from the background surface, particularly in sparse canopy stands, might explain this in open stands, however, it could not account for the differences in IPAR and APAR in the denser canopy, higher LAI stands.

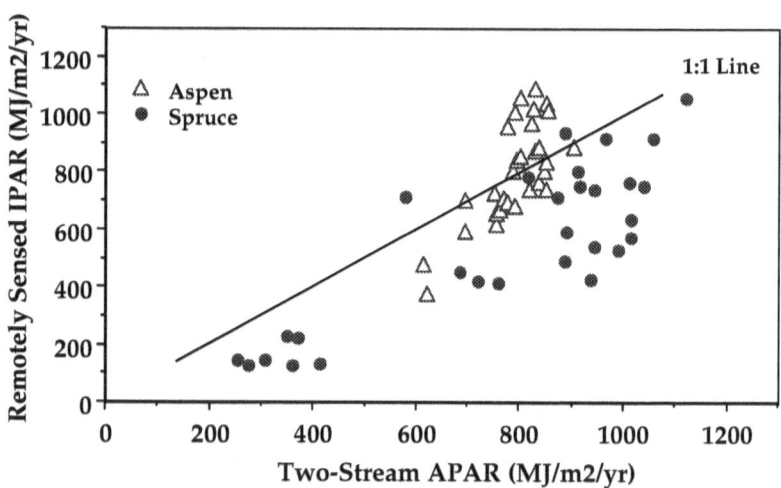

Figure 18. Comparison of simulated annual PAR absorption as calculated from the two-stream model (used in TCX) versus that derived with the remote sensing (geometric-optical) approach.

Rather, the observed differences were most likely due to variations introduced by the strong geometric component of spruce stands, which were characterized as dark shadow-casting objects on a bright background in the par modeling (Chapter II). Simple two-stream calculations of canopy reflectance and light absorption in these canopies were likely to be less accurate than comparable values derived with the geometric-optical approach. Moreover, the geometric-optical approach was better characterized with canopy structural measurements.

Site	APAR (MJ/yr)	APAR Stdev	n (g/MJ)	n Stdev	g (g CO_2/MJ)	g Stdev
2	916	58	0.355	0.024	6.07	0.79
12	256	27	0.154	0.016	3.07	0.64
14	966	59	0.448	0.028	6.69	0.83
15	888	56	0.391	0.025	5.90	0.75
18	371	26	0.170	0.012	3.29	0.46
19	350	19	0.166	0.010	3.27	0.37
38	888	45	0.332	0.018	5.39	0.56
39	580	28	0.203	0.010	4.00	0.39
41	910	54	0.384	0.024	5.96	0.72
42	816	48	0.317	0.019	5.45	0.65
45	944	55	0.379	0.023	5.99	0.71
47	759	36	0.237	0.011	4.52	0.43
48	889	45	0.455	0.024	5.79	0.60
49	1015	53	0.404	0.022	5.99	0.64
50	1014	58	0.426	0.026	6.11	0.72
51	685	42	0.254	0.016	4.76	0.60
52	936	59	0.400	0.027	5.85	0.77
55	944	37	0.382	0.016	5.54	0.44
56	720	35	0.299	0.015	5.68	0.57
57	873	41	0.366	0.018	5.38	0.51
62	307	15	0.164	0.009	3.23	0.33
63	412	19	0.171	0.008	3.33	0.31
64	275	22	0.177	0.015	3.41	0.56
68	990	62	0.386	0.026	6.11	0.80
100	1038	65	0.518	0.034	6.74	0.87
101	1120	62	0.511	0.028	7.75	0.86
102	1009	40	0.343	0.014	5.30	0.42
105	1058	63	0.508	0.032	6.81	0.83

Table 9a. Simulated annual absorbed PAR (APAR), efficiency of net production (n) and of gross production (g) for spruce stands. Units are as indicated in Table 4.

In aspen stands, the relationship between the two-stream APAR and geometric-optical IPAR was unbiased but, as with spruce, it differed between the two approaches. In higher LAI stands the geometric-optical model resulted in higher IPAR estimates than the two-stream

APAR calculations. The opposite was noted in low LAI stands; the geometric-optical model produced somewhat lower estimates than the two-stream approach. As all of the aspen stands had LAI values greater than 1.0, the distinction between intercepted and absorbed PAR was not considered a critical factor in the different results of the two approaches. Rather, measurements of canopy closure were found to be lower for those stands at which APAR (two-stream) was higher than IPAR (geometric-optical) ($r^2 = 0.52$, $p<0.001$). This was consistent with increased background PAR reflectance in sparse canopy stands acting to increase APAR relative to IPAR. Conversely, canopy closure was higher in those stands for which APAR was less than IPAR. Canopy closure measurements, used in the geometric-optical model but not the two-stream calculations, thus accounted for most of the differences in annual PAR harvesting noted between the two approaches for the aspen stands.

Site	APAR (MJ/yr)	APAR Stdev	n (g/MJ)	n Stdev	g (g CO_2/MJ)	g Stdev
3	771.	58	0.730	0.058	5.71	0.88
16	756.	51	0.882	0.059	6.24	0.83
20	755.	54	0.975	0.070	5.74	0.82
21	836.	55	0.706	0.046	6.10	0.80
36	697.	51	0.772	0.058	4.69	0.69
69	792.	58	1.080	0.083	5.22	0.78
71	802.	62	0.980	0.078	4.75	0.74
72	831.	51	1.110	0.069	6.93	0.85
73	851.	48	0.954	0.055	6.90	0.78
74	853.	61	1.110	0.082	6.93	1.01
75	821.	54	1.074	0.072	7.01	0.92
79	874.	50	0.976	0.058	7.23	0.84
80	906.	62	0.972	0.067	6.62	0.91
81	762.	45	0.819	0.049	5.47	0.65
82	824.	48	0.852	0.050	6.41	0.76
83	795.	45	0.858	0.051	5.65	0.66
84	787.	50	0.672	0.044	3.96	0.51
85	802.	47	1.228	0.074	6.93	0.82
86	613.	36	0.502	0.030	3.29	0.39
87	777.	45	1.286	0.075	5.99	0.69
88	853.	50	1.406	0.084	6.55	0.78
89	754.	55	1.176	0.087	5.44	0.80
90	847.	51	1.020	0.061	6.78	0.81
92	837.	60	0.986	0.074	6.91	1.01
93	776.	52	0.884	0.061	6.42	0.88
94	622.	42	0.342	0.023	2.54	0.34
95	696.	51	0.371	0.028	2.78	0.41
96	828.	47	1.039	0.060	6.92	0.79
97	854.	49	1.049	0.060	6.80	0.78
98	829.	49	0.978	0.059	6.77	0.81

Table 9b. Same as Table 9a, but for aspen stands.

In summary, it was clear from both the production efficiency modeling and ecophysiological simulation analyses that there were significant differences in PAR utilization between spruce and aspen, species which are very different with respect to growth rate, life span, deciduous/evergreen habit, and resource allocation. There were also significant differences in $_n$ among age classes of aspen. Some of the sources of this variability were considered in the last chapter; in particular, variations in $_n$ due to differences in below-ground allocation, resource availability, or respiration demands. While each of these causes may result in variability in $_n$, there was not a single independent factor that could conclusively be considered the source of observed variability within the study site. This subject is explored further in the next section.

Variability of PAR Utilization in Relation to R:A Ratio

Because the simulation analyses produced realistic results, and corroborated the remote sensing analyses, the carbon flux simulations with the measured stand variables were used in conjunction with a more generalized set of simulations (i.e., with all variables except LAI and foliage N held constant) to explore sources of variability in $_n$. Figure 19 shows that variability in carbon respiration costs relative to assimilation gains, that is, the R:A ratio or respiration efficiency, was inversely related to variability in $_n$. This was true not only for the stand-specific spruce and aspen simulations, but also for the series of generalized simulations. In spruce stands 82% of the variability in $_n$ was explained by the R:A ratio, and in aspen stands the explained variance ranged from 52% in mature stands to 93% in the young stands. The linear regression relationships of R:A on $_n$ were all significant, as indicated in Table 10. Once again, the lower explained variance in mature stands was a result of the small range in observed values of $_n$ among these stands.

For the simulations with all variables except LAI and foliage N held constant, 92% of variability in $_n$ among spruce stands was explained by the R:A ratio. In aspen stands, the relationship was non-linear, which suggested a saturating effect at maximum LAI and foliage N values. Differences between the stand-specific and the generalized simulations were a result of the otherwise constant parameterization used in the latter (see Figure 13).

The range in respiration efficiency for the generalized spruce simulations had a minimum of ~0.6 and a maximum of ~0.9 (i.e., at least 60% and as much as 90% of the carbon assimilated annually by spruce stands was consumed by respiratory processes). The range in the R:A ratio for the spruce simulations with measured stand variables was smaller, between ~0.7 - 0.8. The range in aspen stands was greater than those for spruce, with a minimum R:A ratio of ~0.30 (70% of assimilated carbon not consumed in respiration) and a maximum of 0.75

(25% not consumed). As with spruce, the aspen simulations based on simulations with measured stand variables were restricted to a narrower range in R:A, between ~0.35 - 0.65, than in the generalized simulations. Thus the simulations based on measured site variables were characterized by a narrower range of respiration efficiencies in both species.

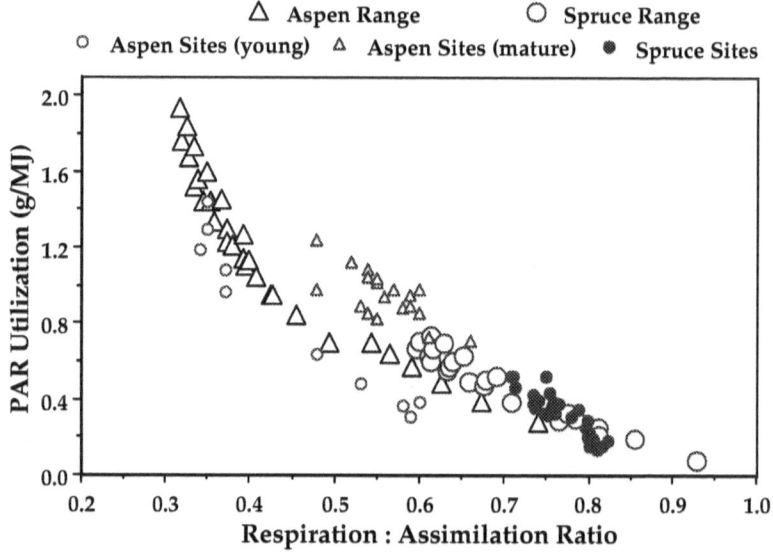

Figure 19. Simulated variation in PAR utilization as a function of variability in the ratio of respiration to assimilation, for both simulations with measured stand variables and for a range of generalized simulations.

The differences in R:A ratio between spruce and aspen were significant (t=15.1, p<0.001). That is, spruce stands required a significantly greater proportion of assimilated carbon for respiratory processes. Young aspen stands had higher efficiencies than the mature stands, which had between ~45 - 65% of assimilated carbon lost to respiratory processes. The differences between age classes of aspen were also significant (t=3.67, p=0.001). This result was consistent with reduced biomass respiration costs in young low-biomass stands, an observation which is further explored later.

Variable	LAI	n	[N]	R:A Ratio	Stand Type
	-	0.88	0.50	0.68	spruce
LAI	-	0.68	0.48	0.63	young aspen
	-	0.09	0.22	0.06	mature aspen
	0.88	-	0.65	0.82	spruce
n	0.68	-	0.93	0.92	young aspen
	0.09	-	0.44	0.52	mature aspen
	0.50	0.65	-	0.37	spruce
[N]	0.48	0.93	-	0.80	young aspen
	0.22	0.44	-	0.01	mature aspen
	0.68	0.82	0.37	-	spruce
R:A	0.63	0.92	0.80	-	young aspen
	0.06	0.52	0.01	-	mature aspen

Table 10. Coefficients of determination on regression relationships for simulations with measured stand variables. significant at 0.005 level. significant at 0.001 level. Matrix is symmetric.

The variation in n and respiration efficiency for the spruce and mature aspen stands is shown in relation to LAI in Figure 20a. Goodness-of-fit statistics are presented in Table 10. Spruce LAI was positively related to n and inversely related to the R:A ratio. In contrast, there was no relationship between LAI and either n or the R:A ratio in mature aspen stands. Young aspen stands were omitted for clarity, but as with the spruce stands, n for the young stands was positively correlated with LAI and inversely related to the R:A ratio. These results suggest that n is not, in some cases, independent of LAI, which is inconsistent with the functional convergence hypothesis. Most of the explanatory power of LAI for n was, however, likely to be the result of a positive correlation between LAI and foliage N content (Table 10). Decoupling the effects of LAI and foliage N is examined in the next section.

Variation in n and respiration efficiency for the stand data is shown in relation to foliage N in Figure 20b. Foliage N was positively related to n in all stands (statistics are provided in Table 10), but was weakly related to the R:A ratio except in young aspen stands. The positive relationship between foliage N and n was not unexpected, owing to greater carboxylation reactions with higher N and associated rubisco concentrations. The poor relationship between foliage N and R:A ratio was, however, an unexpected result as an increase in rubisco might be expected to produce in an increase in assimilation relative to respiration.

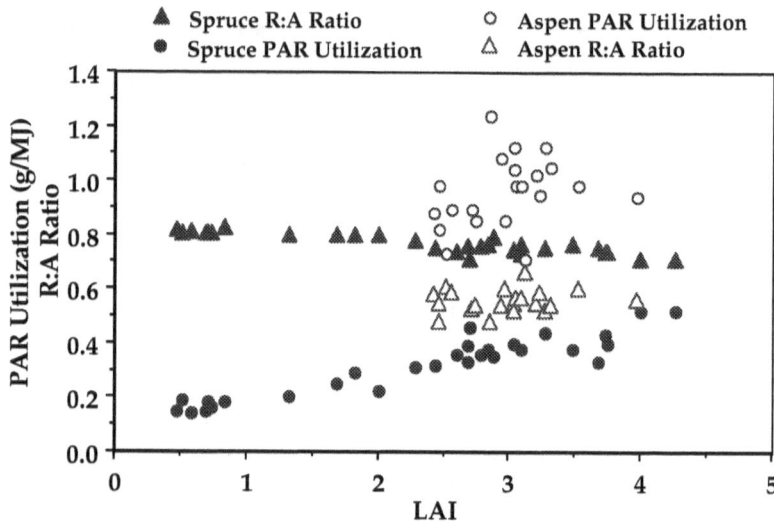

Figure 20a. Simulated variation in PAR utilization and respiration : assimilation ratio, as a function of LAI, for the simulations with measured spruce and aspen stand variables.

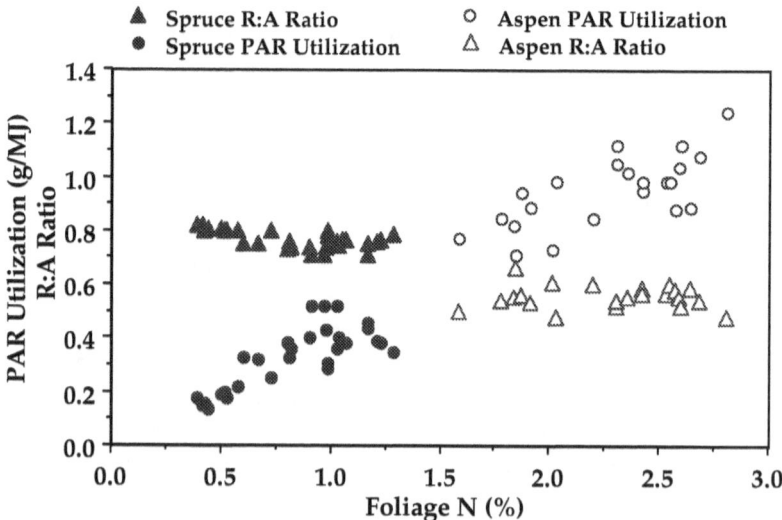

Figure 20b. Same as Figure 20a, but for foliage N content.

Ryan (1995) has shown that respiration per unit dry matter is correlated with foliage N owing to maintenance processes associated with protein turnover. This may account for covariation in assimilation and respiration rates with N, which would limit variability in the R:A ratio. Ryan (1995) also noted, however, no significant correlation between foliage N and respiration rates for *Populus tremuloides*, and he points out that other studies have likewise found poor relationships between the two for individual species. Thus the weak dependence of R:A on foliage N was likely determined by other stand variables, particularly the strong positive correlations between respiring biomass amount (foliage, sapwood and roots) and both respiration and assimilation. This topic is further explored later in the chapter.

With the generalized stand simulations (i.e., all variables except LAI and foliage N held constant) a nonlinear relationship between foliage N and R:A resulted, in both spruce (Figure 21a) and aspen (Figure 21b). The sensitivity of respiration efficiency to foliage N was somewhat greater in spruce, which was the result of a greater rate of increase in assimilation when the lower N values used for the spruce stands were incremented. This is a realistic result in that spruce has been widely noted to be a more N-limited species than aspen. The simulations with measured stand variables were, however, more representative of conditions found in the field than were the general simulations, in that both biomass and respiration tended to increase with increasing assimilation. That is, because biomass was held constant in the general simulations while N varied, greater assimilation was not coupled with increased respiration, thus greater variability in R:A resulted. When extending results from a small number of stands, or a single "representative" stand, to larger areas, model simulations must account for this type of variation in stand biophysical properties.

The generalized spruce simulations resulted in respiration efficiencies approaching the minimum value (i.e., most efficient) near 1.0% foliage N, which was the average N value observed for the stands and which is a typical value observed in the field for the species. A comparable "best" respiration efficiency in aspen was less pronounced, but occurred between 2.0 - 2.5% foliage N, which was again representative of N values typically observed in the field for the species. The lowest respiratory loss of assimilated carbon occurring at the average of the "tuned" foliage N values lends further support to the previously noted evidence that the stand parameterizations were accurate. Moreover, when foliage N was held constant a comparable range of respiration efficiencies emerged from the simulations with measured stand variables (58 - 82% in spruce; 31 - 66% in aspen), and in both cases these were also significantly related to n ($r^2=0.65$ in spruce and $r^2=0.82$ in aspen, p=0.001). These results demonstrate that the dependence of n on respiration efficiency was not a result of varying foliage N as a free variable. Rather it was dependent on the full suite of stand variables listed in Table 6. It is also worth noting that the R:A ratio varied significantly with foliage N only in the young stands,

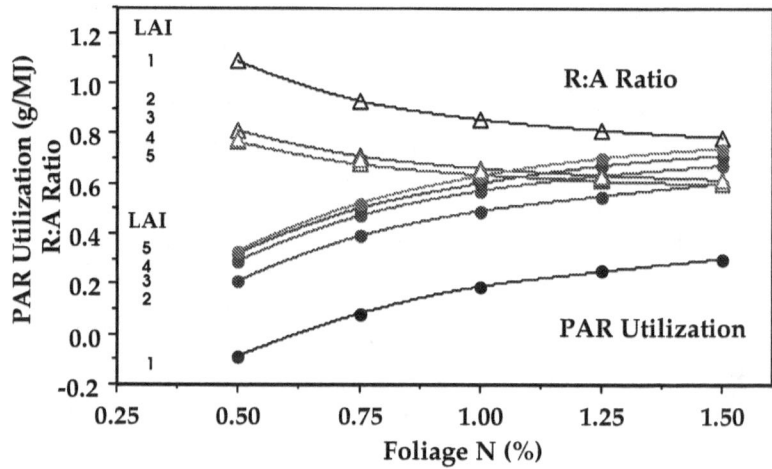

Figure 21a. Result of simulations conducted for spruce stands with all variables except LAI and foliage N held constant.

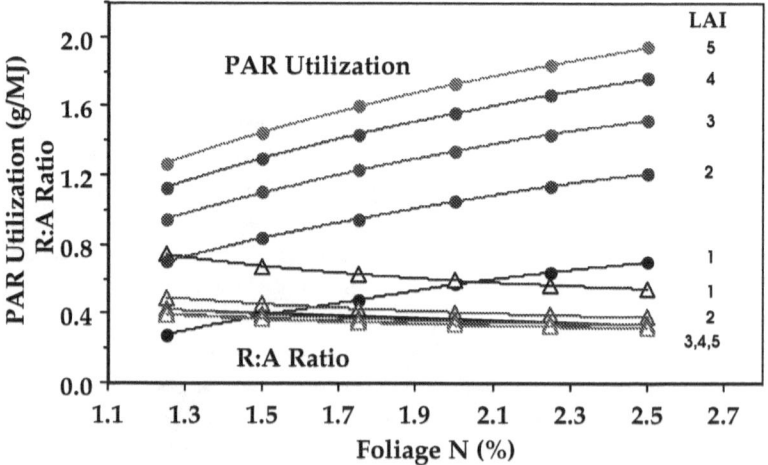

Figure 21b. Same as Figure 21a, but for aspen.

which included stands establishing on poor soils. Such stands may be transient in that they may not reach maturity owing to successional replacement. This suggests that functional convergence operates such that a narrow range of respiration efficiencies result only within stable assemblages of the same species, that is, dynamic equilibrium states reflecting long-term average site conditions. This point is discussed further in the next chapter.

Incorporating LAI into the generalized simulations had no effect on the form of the relationship between either foliage N and n or foliage N and R:A ratio. Instead, LAI modified the simulated values of both variables the same amount at all foliage N concentrations, particularly when LAI was increased from 1.0 to 2.0. Above a value of 2.0, LAI had little effect on simulated values of the R:A ratio and a diminished effect on n, particularly in spruce.

The generalized simulations produced results comparable to the simulations conducted with measured stand data only when the values of foliage N and LAI were restricted to a relatively narrow range of values. For example, a spruce stand shown in the general simulation exercise, with an LAI of 1.0 and a foliage N of 0.5% (and otherwise average biomass allocation variables), produced negative AANPP and n values and an unrealistic R:A value (greater than 1.0). This effect was a result of photosynthetic gains at low LAI being insufficient to offset costs associated with maintaining the metabolic requirements of the plant. These observations further emphasize the validity of the model simulations with the measured stand variables, and they also show that the variation of n with LAI in the simulations with measured stand variables was affected by covariation of foliage N and LAI.

The results thus far have demonstrated not only that there were significant differences in n and respiration efficiency between species, and among stands of the same species, but also that variations in respiration efficiency explained much of the observed variability in n. Moreover, simulations of both the spruce and aspen stands with measured variables, including LAI, biomass allocation, foliage N, *etc.*, resulted in variations in the ratio of respiration to assimilation that were significantly different between species. Higher respiratory costs were noted for the spruce stands, and these were consistent with reported values for *Pinus* stands (Ryan et al. 1994). Presumably this results from a greater utilization of PAR for non-growth processes, a subject which is further explored in the next chapter.

Variability of R:A Ratio in Relation to Biomass

The complement of the R:A ratio is the amount of assimilate remaining after maintenance respiration processes (i.e., $1 - R/A = Y_m$). An empirical model has been proposed that links Y_m with the amount of standing above-ground biomass (Hunt 1994) using established links between maintenance respiration carbon costs and biomass amount (see review, Chapter I and model description, this chapter).

The relationship between above-ground biomass and Y_m, as derived from the simulation results, was not the same for spruce (Figure 22a) and aspen (Figure 22b) stands. Hunt s empirical model produced identical results in both species, as it is based on an exponential decrease in Y_m with biomass amount, independent of species. In both cases, Y_m was invariant with biomass above 50 Mg ha^{-1} in spruce and above ~100 Mg ha^{-1} in aspen. Presumably this reflects a nearly constant proportion of sapwood biomass relative to total above-ground biomass. The simulated values of Y_m in the spruce stands varied little over the entire range of above-ground biomass. Even when the Y_m values were adjusted to reflect just the growing season, to eliminate the effect of winter respiration (shown in Figure 22a), there was little variation in Y_m between stands. Neither was above-ground biomass related to stand age in spruce ($r^2=0.01$; moreover, the two lowest biomass stands were both over 100 years old, whereas one of the highest biomass stands was just 53 years old). Thus, Y_m in spruce stands was nearly invariant with above-ground biomass or age.

Assuming Y_m is dependent on above-ground biomass in the spruce stands would result in an underestimate of respiratory costs and an overestimate of NPP and n. Instead respiratory costs associated with respiring biomass (i.e., sapwood) were balanced by photosynthetic gains, as suggested by a significant relationship between sapwood biomass and assimilation ($r^2=0.91$, $p<0.001$). This also reflects the fact that about half of the growing season respiratory costs of spruce (52%) were related to the maintenance of foliage, which is clearly also linked to assimilation.

In aspen, the simulated values of Y_m were comparable to the results of Hunt s empirical model, although the simulated Y_m values were again systematically lower, and the young stands on poor soils were distinct from other stands. Root biomass allocation of the young stands on poor sites was somewhat higher than at the other young aspen stands (44% versus 37% of total biomass), as would be expected at such sites in the field. Similarly, sapwood amount was somewhat higher on the poor sites (56% versus 45% of above-ground biomass; Figure 12b). These differences in biomass allocation were less likely to be responsible for the observed difference in aspen Y_m than the large difference in average annual soil degree-days at the sites (2053 in the poor sites and 1289 in the other young aspen stands), and the associated differences in mean daily soil moisture content and stomatal resistance (673 versus 457 s m^{-1}) during the growing season. As with spruce, respiratory costs associated with sapwood biomass were significantly related to assimilation ($r^2=0.71$, $p<0.001$). These results explain why biomass was not related to n in the analysis conducted in Chapter II. Even so, the self-thinning nature of aspen stands, characterized by increasing biomass and nearly constant LAI with age, resulted in variability in Y_m in "unstressed" stands that was consistent with the results of Hunt s empirical model.

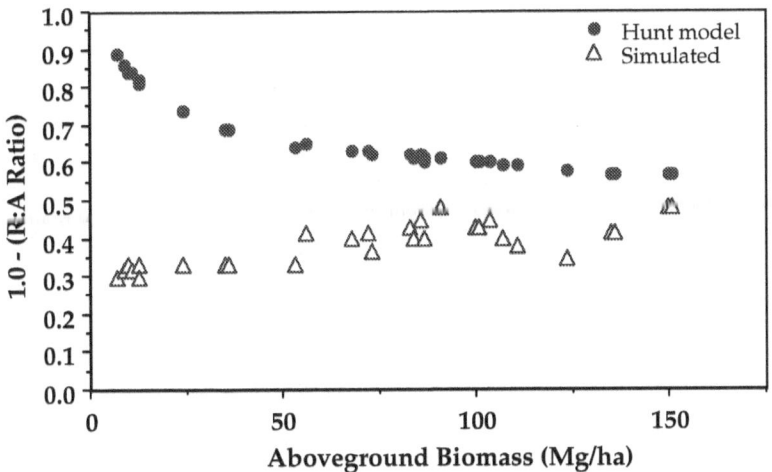

Figure 22a. Variation in the proportion of spruce assimilation not used in maintenance respiration, in relation to standing above-ground biomass.

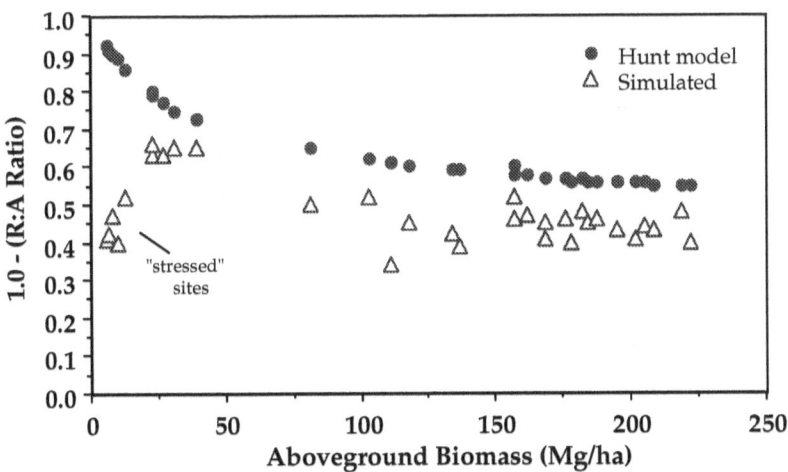

Figure 22b. Same as Figure 22a, but for aspen stands. Stressed sites are discussed in text.

Gross versus Net Carbon Yield of APAR

The final part of the analysis conducted with the ecophysiological model examined which remaining variable(s) may be used to simplify the complexity of ecophysiological models and facilitate the use of light harvesting indices as derived from remote sensing. In addition to PAR harvesting, PAR utilization was identified as an important variable for NPP evaluation, but the results thus far suggest that variability in PAR utilization requires characterization of the R:A ratio. It was not clear if there were variables that remained relatively invariant between species (or plant functional types) despite variations in other individual plant characteristics (e.g., LAI, biomass, *etc.*).

An examination of the stability of total carbon assimilation (rather than NPP) in relation to APAR ($_g$) showed that variability in $_g$ was restricted to a narrower range of values than $_n$. Figure 23 shows the relationship between assimilation and APAR for both the simulations with measured stand variables and the generalized simulations. The range in $_g$ for the simulations with measured stand variables, listed in Table 9a (spruce) and Table 9b (aspen), was once again restricted to a narrower range than the generalized simulations. The young

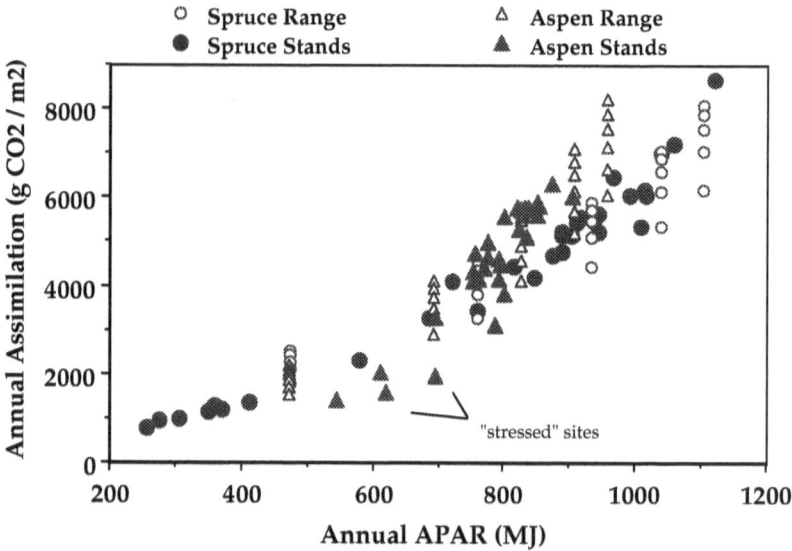

Figure 23. Annual assimilation versus annual APAR ($_g$) for all site-specific (stands) and generalized (range) spruce and aspen simulations. Stressed sites are discussed in text.

aspen stands on poor sites were distinct from the other aspen stands, a result of higher stomatal resistance and lower assimilation rates. The relationship between APAR and assimilation was highly significant (p=0.001) among both the spruce (r^2=0.94) and aspen (r^2=0.85) stands, and the coefficients of variation on g were lower in all cases than those on n (Table 11).

As a result of the reduced variability in g relative to n, the difference in g between spruce and aspen stands was not significant (p>0.14, t=1.62), although the difference between young and mature aspen stands remained significant (p<0.05, t=2.49). The difference in g between the age classes of aspen was largely a result of greater stomatal control on the poorer sites. Thus stomatal control introduced variability in g, which in turn suggests it may limit convergence in g. Additional studies in a range of environments are needed to address this more thoroughly, particularly in areas where water is the primary factor limiting growth.

g	min	max	mean	stdev	CV (%)
Spruce	3.07	7.75	5.214	1.253	24.0
Aspen (Young)	2.54	6.56	4.311	1.481	34.4
Aspen (Mature)	4.69	7.23	6.406	0.653	10.2
Aspen (all stands)	2.54	7.23	5.751	1.378	24.0
n	min	max	mean	stdev	CV (%)
Spruce	0.15	0.52	0.328	0.112	34.1
Aspen (Young)	0.35	1.44	0.825	0.421	51.0
Aspen (Mature)	0.71	1.24	0.949	0.133	14.0
Aspen (all stands)	0.35	1.44	0.910	0.259	28.5

Table 11. Statistics on n and g for stand-specific simulations. CV is coefficient of variation (stdev/mean).

Stand-averaged daily g was very similar for both species (Figure 24), which suggests that differences in g between species on an annual basis were largely due to differences outside of the growing season. For example, those points off the 1:1 line in Figure 24 were periods at the fringe of the growing season when the organic soils of the spruce stands were still frozen but the aspen stands were photosynthetically active. Thus, although average net daily CO_2 flux during the growing season differed significantly between the two species (7.81 versus 12.3 g CO_2 m^{-2} day^{-1}; t=9.78, p<0.001), owing to the higher respiration costs in spruce, mean differences in average daily assimilation per unit APAR (4.72 versus 4.87 g CO_2 MJ^{-1}) were insignificant between these two very different functional types of tree (t=0.34, p>0.5) (see also Figures 16 a,b).

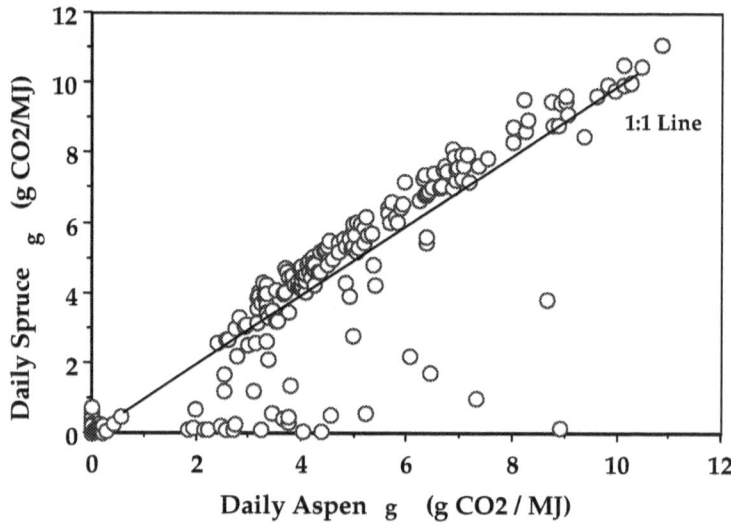

Figure 24. Average daily assimilation per unit APAR ($_g$) in spruce versus aspen stands. Points off the 1:1 line were periods when spruce soil was frozen.

The simulation results suggest that functional convergence is more likely to result in a narrow range in $_g$ than in $_n$ between plant functional types. The results also suggest that functional convergence operates such that a narrow range in the R:A ratio within functional types results. Both of these observations are logical modifications to the functional convergence hypothesis in that $_g$ is closely related to well-established physiological measurements of quantum yield (Prince and Goward 1995), and R:A reflects variations in the benefits of carbon assimliation relative to the respiratory costs of additional biomass production. The generalization of a narrow range in $_g$ and R:A to a broader range of functional types is explored further in the next chapter.

Summary of Simulation Analyses

The simulation modeling has explored the basis for variability in observed values of $_n$. Explicit parameterization of the model for individual stands, and constraining the model to converge on measured values of NPP, allowed comparison with $_n$ values derived from remotely sensed observations. It also allowed examination and utilization of gross and net carbon fluxes associated with assimilation and respiration processes. The stand-specific parameterizations consistently produced a narrower range of results than a series of more generalized simulations conducted with all variables except LAI and foliage N held constant.

Although the production efficiency model (Chapter II) and the ecophysiological model simulations provided somewhat different estimates of light harvesting, owing to different modeling approaches, the simulation results corroborated the remote sensing analyses. Specifically, analyses exploring the relationship between light harvesting and AANPP confirm the distinction between spruce and aspen stands, and the differences associated with age classes in aspen. Results of the simulation analyses for both the spruce and aspen stands were in agreement with other studies that have noted the importance of structure information in the evaluation of canopy light harvesting (Strahler and Jupp 1991; Hall et al. 1995; Huemmrich 1995). Thus, the conclusion reached in Chapter II, that the relationship between annual light harvesting and annual production varied significantly among and between stands in the study area, was confirmed.

The combined results of two different sets of simulations, one with measured stand variables and one with all variables except LAI and foliage N held constant, suggest that the R:A ratio (or the proportion of assimilated carbon lost through respiratory processes, referred to as respiration efficiency) remains largely invariant with changes in foliage N and LAI, whereas n varies strongly with foliage N, and to a much lesser extent, with LAI. There was a clear distinction in respiration efficiency between spruce and aspen stands, but small variations within stands of the same species. The only exception to this was significant variation in R:A with foliage N in young aspen stands, which were separable into low productivity stands on shallow soils (i.e., higher moisture stress) and higher productivity stands on better sites. Stressed stands incur greater carbon costs associated with establishment on poor sites and may not be competitively viable in the longer-term. Thus, within-species convergence in respiration efficiency on a narrow range of values was restricted to stable vegetation assemblages.

Results of the modeling work also established that much of the variability in n can be attributed to the significant difference in respiration efficiency between spruce and aspen, and the smaller but significant variability in respiration efficiency within aspen. The link between respiration efficiency and n in the simulations with measured stand variables was attributed to the combined response of the measured variables (i.e., LAI, biomass allocation, foliage N, soil properties, *etc*: Table 6), and parameters associated with vegetation functional type (respiration coefficients and carbon partitioning; Table 7). Thus, the combined effect of all the variables required to simulate stands found in nature acted to reduce variability in R:A, which, in turn, accounted for much of the variability in n. Moreover, parameterization of the individual stands with foliage N held constant still resulted in a range in R:A that was sufficient to explain a comparable and significant amount of the variability in n.

The complement of the R:A ratio (Y_m) in spruce stands was shown to be largely invariant with standing above-ground biomass amount, which was not consistent with the results of an empirical model developed by Hunt (1994) to estimate Y_m. This result was attributed to the covariation of assimilation and respiration in the simulations with measured stand variables, and the suite of life history adaptations of spruce (e.g., slow growing, long-lived). In particular, there was a constant proportion of respiring sapwood relative to total woody biomass, and a large proportion of Y_m was from foliage (rather than woody) biomass. Together these resulted in respiratory losses being balanced by concurrent photosynthetic gains, thus in nearly constant Y_m. The simulation results for Y_m were in better agreement with Hunt s empirical model for aspen stands, which was largely a result of proportionately lower respiratory costs in young (low biomass) stands. Young stands on poor soils were, however, distinct from all other aspen stands owing to reduced assimilation associated with moisture stress and stomatal control. As noted, such stands may not be competitively viable.

The observed insensitivity of Y_m to above-ground biomass suggests that it is necessary to consider plant functional type differences in evaluating respiratory costs. It may be possible to approach this problem through estimates of the ratio of foliage to total biomass, where foliage biomass were retrieved through inversion of spectral vegetation index (SVI) - LAI relationships. This would require work outside the realm of this dissertation.

In contrast to significant differences in seasonally averaged net CO_2 exchange and n between species, there was an insignificant difference in g, the amount of carbon assimilated per unit APAR. This result lends support to an assumption of the PEM model (reviewed in Chapter I) that g is related to the quantum yield, which measurements have shown to be relatively invariant. Greater moisture stress in a few aspen stands on poor soils, however, induced greater stomatal control and resulted in a significant difference in g from the other aspen stands. Thus stomatal control may act to diminish the convergence in g. This subject is explored further in the next chapter.

The results of the simulation modeling analyses were consistent with what would be expected as a result of functional convergence. In particular, this includes the following findings: (i) a constrained range of conditions exhibited by the simulations with measured stand variables relative to the more generalized simulations; (ii) a narrow range of R:A within stands of the same species; (iii) an apparent convergence on g rather than n between species. As a result, modification of the functional convergence hypothesis, as it was originally proposed, appears to be necessary. This subject, and the possibility of generalizing the conclusions reached here, is explored in the next, and final, chapter.

84

Chapter IV. Analysis of Results in the Context of Functional Convergence

This final chapter of the dissertation focuses on the implications of the previous chapter results for NPP modeling, and their generalization to a wider range of conditions and vegetation types to those found in boreal forests. Specifically, it reintroduces the functional convergence hypothesis (Field 1991), elaborates its relevance to NPP modeling in general, and then assesses its utility to satellite remote sensing of NPP in particular. The research conducted as part of the analysis of functional convergence were of a more conceptual nature than the experimental approaches used in the previous chapters. Thus it largely entails reasoning rather than experimentation, with the goal of testing the hypothesis than functional convergence results in a narrow range of n.

In brief, the functional convergence hypothesis states that resource availability and resource acquisition costs result in an optimization of resource-use efficiency by plants through evolutionary selection, which in turn results in a maximization of carbon gain and a narrow range of n values (Field and Mooney 1986; Field 1991). The hypothesis is supported by field measurements of net ecosystem CO_2 exchange, which have been shown to be remarkably similar, as a function of incident PAR, among tropical, temperate and boreal forests (Fan et al. 1990). The results up to this point suggest, however, that functional convergence does not result in a narrow range of n in the forest stands examined here, or in many other studies of various plant functional types.

The hypothesis is worth examining further, nonetheless, as it has a basis in powerful arguments rooted in evolutionary biology. Moreover, the model results showed a narrow range in R:A within species and in g among species (or functional types), which suggests convergence may be reflected in these variables rather than in n owing to different carbon costs among and between functional types of plants. Assessing generalization of these results requires consideration of the evidence for which net photosynthesis can be regarded as the result of optimized resource utilization in the context of limited resources.

Resource Constraints and Adaptive Strategies

Adaptive strategies, defined as suites of co-adapted characteristics, are thought to arise as natural selection acts on a range of traits in a population. Strategies that are stable in an evolutionary sense, or have a "dynamic stability" are those most likely to persist (Maynard Smith 1976). There is ample evidence that tradeoffs in resource allocation result in suites of life history traits (Gadgil and Bossert 1970; Stearns 1976, 1989; Caswell 1982; Ricklefs 1991). These suites of traits have been classified in a number of ways, including MacArthur and Wilson s (1967) widely applied model of "r" versus "K" selection in island bird species, Grime s (1977) ternary model of primary plant growth strategies, and Tilman s (1988) model of relative growth rate (RGR) and biomass partitioning. There have been recurring debates on the validity and utility of these classification strategies (Shipley and Peters 1990; Tilman 1991; Silvertown et al. 1992) but their existence is not in doubt.

For example, studies of many species at the leaf, plant, stand and ecosystem level demonstrate that the life-span of a plant s foliage is closely related to other traits in the suite, including photosynthetic capacity ($_c$), whole-plant life-span, and RGR (Lamber and Poorter 1992; Reich et al. 1992). The relation between $_c$ and RGR has long been noted in needleleaf evergreen versus broadleaf deciduous trees grown under similar conditions and is documented in data compiled for 159 tree species of North America (Loehle 1988). Other associations between life history traits have been reported in a number of studies (Stearns 1989; Ricklefs 1991; Chapin 1993).

Although relationships among traits of a given suite may not all be tightly linked, a consensus has emerged that the association of traits is a result of their co-evolution (Partridge and Harvey 1988). An example of trait relatedness in fast- versus slow-growing tree species is shown in Figure 25. Notable exceptions to these generalizations may occur as plants age or change successional status (Lerdau 1992; Gleeson and Tilman 1994). For example, there is evidence that plant size influences linkages between traits (Shipley and Peters 1990; Chapin 1993). Thus the trait relatedness expressed in different growth-forms is determined by dynamic resource allocation strategies which may change with ontogeny or changes in stability of the environment (Kawecki 1993; Perrin and Sibley 1993).

Allocation Strategies, Defenses Costs and Payback Intervals

It has been suggested that allocation of assimilated carbon in trees follows a hierarchy, where the first priority is given to maintenance respiration of the living tissues, next to production of foliage and fine roots, then to flower and seed production (which may slow growth of other components), then to primary growth (terminal and lateral branch growth and root extension), and finally to addition of xylem and defensive compounds (Oliver and Larson 1990). This resource partitioning results in enhanced allocation to those components associated with extraction of limiting resources. One of the characteristics of fast growing species is that they have a greater ability to adjust allocation to different plant components, for example in response to stress (Kachi and Rorison 1989; Van der Werf et al. 1993; Laurence et al. 1994). Gleeson and Tilman (1994) suggest "observed allocation patterns may be better viewed as adaptively plastic responses to resource gradients." This plasticity is particularly evident in specific leaf area (Lambers and Poorter 1992).

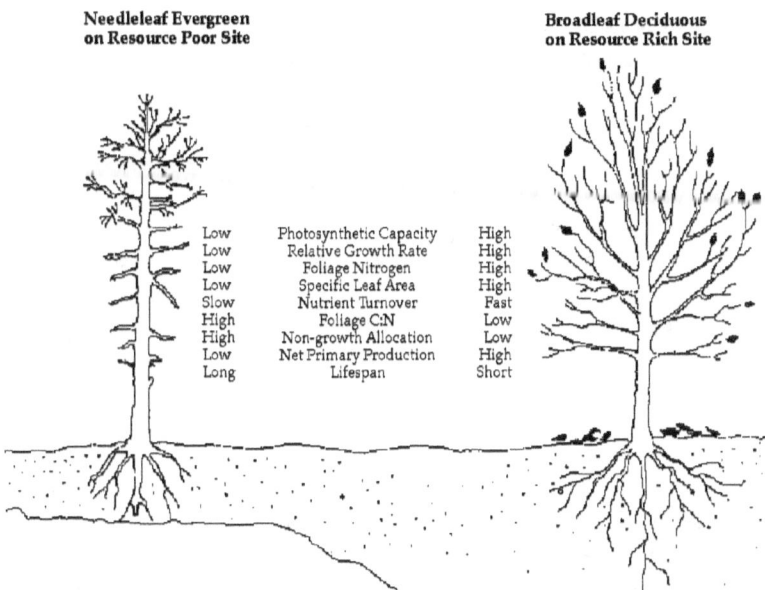

**Needleleaf Evergreen
on Resource Poor Site**

**Broadleaf Deciduous
on Resource Rich Site**

Low	Photosynthetic Capacity	High
Low	Relative Growth Rate	High
Low	Foliage Nitrogen	High
Low	Specific Leaf Area	High
Slow	Nutrient Turnover	Fast
High	Foliage C:N	Low
High	Non-growth Allocation	Low
Low	Net Primary Production	High
Long	Lifespan	Short

Figure 25. Generalized suites of life-history traits associated with fast-growing, short-lived tree species in resource-rich sites and slow-growing, long-lived species in resource-poor sites.

Differences in allocation strategies have also been linked to increased relative investment in supporting structures and secondary compounds required for defense against herbivores and pathogens (Coley et al. 1985; Field and Mooney 1986; Fagerstr m et al. 1987; Loehle 1988; Lerdau 1992). This subject is an important consideration in the study of resource utilization and its relation to NPP because costs of assimilate allocation to non-photosynthetic components are not negligible and vary among functional types. Allocation to defenses can range from 10 to 30% of dry weight in herbaceous species and even more in woody species (Coley et al. 1985; Coley 1986).

Chemical defenses (including resins and phenolics such as lignin and tannins) are of particular significance in resource utilization because they require a higher allocation of assimilate, and have higher respiration costs for synthesis (Griffin 1994). For example, the CO_2 release associated with the synthesis of defensive chemicals are 9 to 15 times that associated with sucrose synthesis and 4 to 10 times that associated with cellulose synthesis (Penning de Vries 1975). CO_2 release associated with the synthesis of volatile terpenoids related to leaf protection in high radiation regimes is even greater.

Economic analogies have frequently been applied to describe these allocations in terms of tradeoffs between costs and returns (Bloom et al. 1985; Chapin 1989; Lerdau 1992). Although the costs versus the benefits of defense processes are difficult to quantify (Raven 1986), observed negative correlation between life-span and RGR have been attributed to the high energetic cost of defensive chemical synthesis and the diversion of assimilated carbon from current growth (Chapin et al. 1987; Lambers and Poorter 1992; Lerdau 1992). Loehle (1988) showed that the allocation to defense in relation to growth rate and longevity could account for observed differences in tree growth rates through mechanisms by which species either defend themselves (slow-growing, long-lived) or outgrow pathogens (fast-growing, short-lived). Similar correlations have been noted at the leaf level (Chabot and Hicks 1982; Coley 1986; Harper 1989; Williams et al. 1989).

Higher relative growth rates and lower defense allocation have been related to lower respiration rates relative to carbon assimilation in fast-growing species (Lambers and Rychter 1989). Lambers and Poorter (1992) suggest this is due to a less energy (carbon) demanding metabolic pathway for nitrogen acquisition in fast compared with slow-growing species. This is consistent with Loehle s (1988) observation of increased respiration costs associated with non-growth processes, and suggests differences in respiration efficiency (the ratio of respiration to assimilation) between functional types. At least one study has demonstrated a positive correlation between measured respiration efficiency and relative growth rate (Anekonda et al. 1994). Several others have established links between what can be considered surrogates of respiration efficiency (cost versus benefit of foliage construction and maintenance) and plant carbon dynamics (Williams et al. 1989; Sobrado 1991; Griffin 1994; Sobrado 1994). For example, mean maintenance costs of deciduous (faster growing) species in a tropical dry forest ecosystem were less than half those of evergreen (slower growing) species (0.82 ± 0.35 and 2.09 ± 0.67 g glucose m^{-2} day^{-1} respectively, Sobrado 1991). Moreover, there is evidence that resource use efficiencies decrease with increasing construction and maintenance costs (Sobrado 1991; Griffin 1994) but that these increased costs are eventually recovered through greater foliage longevity (i.e., longer payback period) (Williams 1989; Griffin 1994).

Figure 26 shows the relative costs of constructing foliage for a range of different plant functional types. Because construction and maintenance costs are highly correlated (Sobrado 1991, 1994) the construction costs shown in Figure 26 can be considered representative of maintenance costs as well. Note that for different plant growth forms (e.g., shrubs, trees) in the same ecosystems (e.g., tropical dry forest, Mediterranean) evergreen species have higher construction and maintenance costs than deciduous species. This observation is consistent with decreased respiration efficiency and increased payback intervals in evergreen (greater foliage longevity) species. Respiration efficiency also varies within functional types. For example, a survey of the literature for complete stand carbon budgets of *Pinus* species, found in a number of different environments and observed under a variety of conditions, resulted in respiration efficiencies between 0.32 and 0.64 (N=7) (Ryan et al. 1994). Whereas the results from Chapter II suggest that the variability within functional types is smaller than between functional types, the range noted by Ryan et al. is not inconsequential.

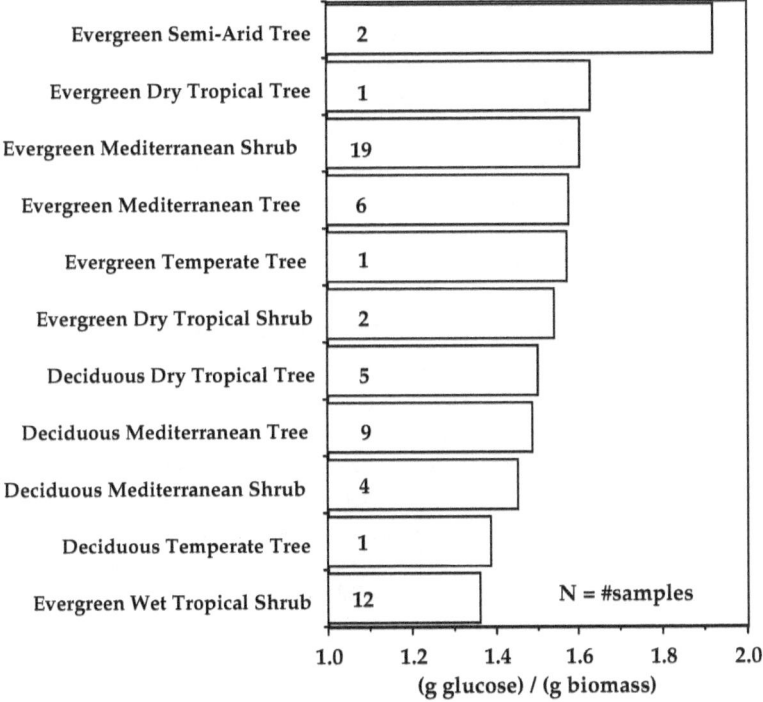

Figure 26. Foliage construction costs of various plant functional types (data from Griffin 1994). Maintenance costs are proportional (see text). Numbers on bars indicate sample size.

The observed variability in respiration efficiency suggest that the likelihood of a nearly constant values among functional types is small, despite the fact that respiration and total assimilation are clearly correlated (see Chapter III; Amthor 1995). Suggestions of a constant respiration efficiency (at ~0.4) (Gifford 1994) are not likely to extend outside of controlled environmental conditions. This conclusion is also supported by extensive observations that maintenance respiration is better related to respiring mass than to total assimilation (e.g., Biscoe et al. 1975; Ryan 1990).

Thus a substantial body of evidence links suites of life history traits with costs and benefits associated with resource acquisition and allocation. Owing to the linkages of traits in adaptive strategies, several plant variables can be used as diagnostic characteristics or "ecological integrators". Vitousek (1982) and Field and Mooney (1986) view nitrogen utilization as one such integrator. Reich et al. (1992) identify foliage longevity as an integrator, "one of several inter-related traits that cannot be viewed in isolation." Chapin (1993) regards RGR as the "central figure of plant adaptive strategies," which allows it to be used "as an index of other features."

The common theme shared by these studies is that allocation strategies and life-history traits reflect an efficient use of available resources. Some have suggested the suites of selected traits have resulted in an optimization of resource use such that fitness is maximized (Maynard Smith 1978). Principles of natural selection have even been applied to computer optimization techniques (Forrest 1993). In evolutionary biology, this "optimality theory" has become a long-running point of debate, beginning with an initial refutation by Gould and Lewontin (1978) who suggested that natural selection simply reflects phylogenetic history and not necessarily an optimized design (reviewed by Parker and Maynard Smith 1990; Dennett 1995). The functional convergence hypothesis explored here has its roots in optimality theory, and thereby requires an assessment of the evidence for resource-use optimization in plants.

Evidence for Optimization of Resource Allocation and Use Efficiency

In conditions of sub-optimal resource supply, two types of response might be elicited in plants, one to increase the allocation of existing resources to the capture of the resource in short supply and the other to increase the efficiency of use of the resource in short supply. An overview follows of the allocation and use efficiencies of resources when in sub-optimal supply considering water, nitrogen and PAR.

Water

The trade-offs involved in maximizing CO_2 uptake under conditions of water stress have been widely studied (Cowan 1986; Givinish 1986) and expressed in a number of indices (Larcher 1995). Of these indices, water use efficiency (WUE) is the most relevant here expressing, as it does, the ratio of dry matter production to water consumption by the plant or stand.

Increased WUE in conditions of water shortage is well-documented as are the differences between species and plants having different carbon fixation pathways (i.e., C3, C4, CAM) (Larcher 1995). Mechanisms have been identified at all temporal scales from minutes to evolutionary time scales. Over long time-periods the emergence of different carbon fixation pathways and their associated physiological and anatomical characteristics can be interpreted as an optimization of carbon fixation per unit of available water. This gives rise to the well-known differences in WUE between these functional types of plant. For example, Abrams et al. (1994) link WUE and leaf structural properties (leaf thickness and mass per unit area) to leaf CO_2 exchange in a range of temperate tree species.

At the time-scale of days plants are known to reallocate carbon to the synthesis of osmotically active substances, the synthesis of abscisic acid (ABA) which has a direct action on stomatal guard cells and on carbon allocation to leaf growth, and to fine root growth(Geiger and Servaites 1991). At the time scale of minutes adjustments of leaf attitude, orientation, folding and curling may reduce the net radiation load (Forseth and Ehleringer 1982; Ehleringer and Werk 1986) thereby reducing water loss relative to carbon dioxide uptake and so increasing the WUE.

Nitrogen

Hypotheses of optimized benefits relative to investment costs for nitrogen acquisition have been explored for more than a decade (Gutschick 1981; Field 1983). Vitousek (1982) provided evidence that nutrient use efficiency (carbon gain relative to nutrient use) in a range of mature forest ecosystems is inversely related to nutrient availability. Two mechanisms were advanced to explain this observation; in nutrient poor sites there is increased resorption of nutrients from leaves to stems prior to abscission, and increased nitrogen use efficiency in photosynthetically active leaves.

There has been a plethora of studies that confirm Vitousek s observations. It has been suggested that because resource allocation to construction of new foliage and uptake mechanisms is expensive for slow growing species, high nutrient use efficiency is positively related to foliage longevity (Field 1983; Reich et al. 1992), the retention of nutrients in more productive foliage (Hom and Oechel 1983), efficient nutrient resorption before shedding (Pugnaire and Chapin 1992), and reduced costs of nutrient storage (Chapin et al. 1990; Monson et

91

al. 1994). For example, in the foliage of black spruce stands in Alaska (a slow growing "stress tolerant" species), nitrogen content of new foliage declined to 70% of maximum values in the oldest foliage (which was 13 years old) (Hom and Oechel 1983). A similar trend was observed with phosphorus, which declined to 55% of maximum values in the oldest foliage.

The most significant aspect of these studies for the present argument is that there is a dynamic optimization of allocation of N in plant tissues when resources are limiting (Bryant et al. 1983; Field and Mooney 1986; Hilbert 1990; Gleeson 1993; Gutschick 1993; van der Werf et al. 1993). This dynamic process has also been incorporated into general ecophysiological models such that critical C:N ratios are maintained (Parton et al. 1988; McGuire et al. 1992), including the TCX model used in Chapter III (Bonan 1993a).

Photosynthetically Active Radiation (PAR)

Efficiency of PAR utilization was a focus of this research and requires assessment in the context of availability and utilization of other resources. It is worth noting, however, a few fundamental aspects of light utilization at this point. Optimization of light capture relative to costs is well illustrated by the evidence for resource "foraging" (Grime 1979; Givnish 1982) by means of increased leaf or height growth, more favorable positioning of leaves, and changes in specific leaf area all of which are associated with species that typically occupy shaded sites. Species that occupy sites in which the light environment may change rapidly from full sun to partial shade typically have enhanced capability of modification of all these structural variables (Bjrkman 1981; Walters and Field 1987; Ellsworth and Reich 1992). In contrast to allocation and structural adjustments, light use efficiencies can change through time, mainly through changes in respiration with ontogeny (Ryan and Waring 1992; Perrin and Sibley 1993) or due to environmental physiology (Potter et al., 1993; Waring et al. 1995) (and shown in the results of Chapters II and III).

Combined Resource Use Efficiencies

The value of n results from the response of plants to a combination of limited resources. Various aspects of light use efficiency are linked to the availability of other resources such that a combined response is elicited to any single resource that limits NPP, reproduction and growth. The combined responses of plants to multiple limited resources enhances a simple "law of the minimum" concept (Bloom et al. 1985; Chapin et al. 1987). For example, the distribution of N and the amount of PAR absorption within *Solidago altissima* canopies has been shown to co-vary (Hirose and Werger 1987). Both PAR and nitrogen concentration per unit leaf area decreased exponentially through the canopy. Reallocation of nitrogen to the sunlit portion of the canopy with foliage aging resulted in maintenance of high c and resource use efficiency.

This relationship has been noted in a number of studies in a diverse array of habitats (Field 1983; Hollinger 1989; Chen et al. 1993; Ellsworth and Reich 1993; Schmid and Bazzaz 1994). Moreover it has been mathematically formalized in the "time mean radiation weighted profile of canopy PAR," which scales a PAR use parameter of canopy performance at all levels of a canopy to that of the uppermost illuminated leaves (Sellers et al. 1992).

Thus resource allocation patterns, when combined with locally adapted canopy geometry (Horn 1971; Norman and Jarvis 1974; Hutchinson et al. 1986), act to efficiently utilize the canopy light environment in relation to other resource levels. In this way resource use is scaled to the limits of other resources available to the plant in a coordinated fashion (see Chapin et al. 1987; Mooney et al. 1991). Moreover, in a review of plant response to multiple stresses, Lambers and Poorter (1992) found little evidence for trade-offs in resource-use efficiencies; the efficient utilization of one resource was not compromised at the expense of the efficient utilization other resources. Thus optimization of resource use efficiency could result in a narrow range of n, as the functional convergence hypothesis predicts.

Links Between Optimization and Leaf Mass Per Unit Area

Recent work suggests that the expression of energy use on a mass-basis, as is the case in n, is the most appropriate basis on which to assess resource-use optimization. Field and Mooney (1986), Reich et al. (1992), and Agust et al. (1994), compared a wide variety of plant species that occur in a diverse range of environments, and noted that resource use efficiencies converge on a leaf mass basis, not on a leaf area basis. Reich et al. (1992) found NPP per unit foliage biomass was closely related to foliage longevity ($r^2 = 0.78$), and varied by almost an order of magnitude. In contrast NPP/LAI was not significantly correlated with either leaf life-span or $_c$. Field and Mooney (1986) noted the relationship between foliar nitrogen concentration and net photosynthesis was nearly linear when expressed on a leaf-mass basis, yet relatively poorly related on a leaf-area basis. Furthermore there was no apparent maximum in the relationship; net photosynthesis showed no indication of leveling off as long as there was sufficient nitrogen available for production of light harvesting machinery and carboxylation enzymes.

A mass basis for resource-use optimization indicates a direct relationship between expenditure of captured resources by the plant and environmental resource distribution. Links between the canopy distribution of resources and the distribution of plant mass reflect the cost of acquisition, whereas areal metrics merely reflect the absolute acquisition of resources. Thus the distribution of resources is linked with the distribution of leaf mass per unit area (LMA, $g\ m^{-2}$). The previously discussed evidence for efficient canopy N distribution has been shown to reflect this mass distribution, that is, nitrogen is distributed both within and among canopies (and light environment habitats) according to a scaling of LMA rather than leaf area (Walters and Field 1987; Gutschik and Wiegel 1988; Ellsworth and Reich 1993).

Thus, available evidence suggests that resource-use optimization, resulting from combined resource-use efficiencies, is apparent when the distribution of those resources is considered in the context of LMA. Moreover, resource use efficiency and LMA are correlated and together determine carbon assimilation. The significance of this in the consideration of n and its application to remote sensing of NPP lies in the links between light harvesting, LMA and canopy reflectance.

Optimized or Compromised Resource Use?

Some researchers have preferred to refer to the evidence for optimization as "coordination" of resources (Chen et al. 1993). Others have discussed it in the context of multiple evolutionary stable strategies that get as close to an optimal solution as genetic mechanisms permit (infinite time and infinite populations would be required to achieve the theoretical optimum) (Parker and Maynard Smith 1990). This debate reflects increasingly abundant evidence that the established linkages between traits, the trade-off costs associated with different allocation strategies, and the response of plants to a variety of limited resources all provide support to the idea that plant evolution has resulted in optimized resource use. A corollary to the evidence for optimization is that "wasteful allocation," such as unnecessary light harvesting and carbon-fixation machinery (defined in terms of the most limiting resource that prevents utilization of additional carbon gain) is selected against.

It has been suggested that "natural selection can be expected to favor not a maximum gross benefit but a compromise, optimum gross benefit which maximizes the net benefit" (Begon et al. 1990). This compromise solution encompasses the available evidence for resource use efficiency, and is exemplified in Reich et al. s (1992) summary of the association between efficient allocation strategies and resulting resource utilization of slow- versus fast-growing tree species (Figure 25). High leaf N concentrations result in higher CO_2 assimilation and greater canopy growth in proportion to total plant growth, which in turn results in greater canopy expansion and height growth and increased canopy carbon gain. Increased carbon gain results in high leaf turnover rates (short leaf life-spans), and associated rapid nutrient recycling. Slow growing species are the reverse. Although this view is simplified, it can be considered a general example of the mechanism by which resource use efficiency, integrated through time, results in a maximized net benefit for different functional types.

94

Links Between Maximization of Fitness and Carbon Gain

Natural selection suggests that resource-use optimization results in maximization of fitness in terms of both survival and lifetime reproductive potential (Doust 1989; Sibley 1989; Burns 1992; Perrin and Sibley 1993). Does this translate into a maximization of carbon gain, in terms of either NPP (biomass accumulation) or gross primary productivity (total CO_2 assimilation), or is fitness unrelated to carbon gain?

The different allocation of resources to various plant processes, and the different respiration efficiencies associated with allocation strategies that are observed suggest that the trade offs between resource allocation for growth versus non-growth (secondary) processes will determine the fate of assimilated carbon. Assimilation needs to be sufficient to provide the organic sugars for the various metabolic requirements of the plant, but that energy is not directly translated into biomass since other processes (such as defense) make different energy demands from respiration of fixed carbon.

Maximization of carbon gain should confer fitness by providing the energy that can be allocated in any of a number of strategies that in turn maximize competitive advantage. Genotypes that assimilate, allocate and utilize resources more efficiently than others that compete for the same resources generally would prevail, resulting in a causal relationship between optimized resource use, maximized carbon gain, and fitness (Parker and Smith 1990; Sibley 1991; Burns 1992). This observation assumes a long-term selection process under conditions of dynamic equilibrium, that is, under transient conditions in which vegetation may change with climate, environmental conditions, disturbance frequency, and so on. Thus, the competitive advantage of a species may be lost as conditions change or are reinitialized by a disturbance event, and this is in turn linked to the life history (including life span) of the species (e.g., Ehleringer 1993; Donovan and Ehleringer 1994). Similarly, competitive advantage might only be inferred by more efficient resource use if this included the combined utilization of all resources, rather than a single resource (whether nutrients, water or light).

A mixture of adaptive strategies within an ecosystem may provide the means by which a maximization of carbon gain is achieved at the level of the plot, stand or ecosystem rather than on an individual plant basis. That is, a variety of solutions by a stand of competing individuals can result in maximized carbon gain, scaled to the available resources (Maynard Smith 1978; Field 1991; Schmidt and Bazaaz 1994).

Although optimization of resource-use efficiency within available resource constraints may result in a maximization of carbon gain, this will not necessarily result in a maximization of biomass accumulation unless such allocation would somehow confer fitness or reproductive potential. There is no direct evidence that accumulation of biomass should be related to either of these, and a maximization of NPP does not necessarily follow owing to a differential in the energy costs associated with various growth and allocation strategies. This is apparently true even within the same species. A study of bigtooth aspen (*Populus grandidentata*) stands over a range of site qualities demonstrated that net CO_2 assimilation was positively correlated with above-ground production, but was not its sole determinant (Briggs et al. 1986). A review by Korner (1991) describes additional cases where net photosynthesis may not provide a good prediction of NPP, and the findings of several studies with a large number of species (reviewed earlier) support this conclusion.

Thus, theory and available evidence concur in supporting the view that convergence of values of $_g$ but not $_n$ may occur. This was further supported by the modeling results (Chapter III), which showed that CO_2 assimilation was not significantly different, as a function of incident PAR, between two distinct functional types represented by stands of short-lived, fast-growing aspen and long-lived, slow-growing spruce. In contrast to $_g$, there is, at best, weak support for a convergence on $_n$ owing to differences in respiration efficiency associated with growth versus non-growth processes and associated payback period in different plant functional types. In brief, the link between fitness and gross carbon assimilation is direct and the link between fitness and NPP is less direct, although the two are likely to be positively correlated through their common link with maximized carbon assimilation.

Implications of Functional Convergence for Modeling Net Primary Production

The evidence presented thus far suggests that many of the variables discussed are not independent, thus ecophysiological models that incorporate several of these may be functionally simpler than they are stated. This has been demonstrated in at least one recent exercise utilizing remotely sensed data and an ecophysiological model (Bonan 1993b) and provides credibility to models that simplify vegetation into spatially uniform variables (characterized as "big leaf" approaches). Moreover, the most sensitive components of the models are frequently those which can be provided by remote sensing, that is, foliage dynamics (amount and longevity) and prominent physiological and morphological traits reflected in life history classifications (e.g., $_c$, relative growth rates, specific leaf area, *etc.*).

The dependence of net photosynthesis and production efficiency on the mass of a plant, as opposed to the area it occupies, is an important consideration to this argument because spectral vegetation indices (SVIs) derived from remote sensing have been linked almost exclusively to leaf area rather than leaf mass. Recent evidence suggests that leaf mass, leaf optical properties and PAR absorption are all highly correlated. For example, Agust et al. (1994) found strong statistical relationships between the amount of light captured per unit mass of photosynthetic tissue, and both the chlorophyll concentration (r^2=0.87) and optical properties (r^2=0.74) of those tissues. Thus remote sensing may be linked to the suite of plant physiological traits through their common dependence on LMA.

Another important consideration for both the remote sensing and ecophysiological modeling approaches that was revealed is the differential in energy costs associated with construction of different plant components, and whether they are for growth or non-growth purposes. Because respiration costs are different for these processes, the use of PAR absorption with remote sensing to estimate NPP requires augmentation with some estimate of relative respiration costs, as was also suggested by the results of Chapters II and III. This factor must be accounted for with respiration terms incorporated into NPP models driven with light harvesting, such as those suggested for the production efficiency model reviewed in Chapter I.

A final factor that requires consideration for evaluating NPP with remote sensing is the coupling of vegetation canopies with the atmosphere. This coupling, described by the "omega factor" of McNaughton and Jarvis (1991), affects the strength of the relationship between canopy PAR absorption and canopy carbon assimilation through a combination of stomatal and boundary layer conductances at a variety of spatial scales. Canopies that are more closely coupled with the atmosphere may exhibit short-term divergence between incident PAR, resource-based foliage display and net canopy carbon gain (hence NPP). This is more likely to be the case with forests (higher aerodynamic roughness and greater boundary layer conductance) than with grasslands or crops (Verma et al. 1986), but may also change with phenological stage (Valentini et al. 1995). Evidence for decoupling of light harvesting and utilization has been presented in Chapters II and III. Thus short-term physiological control of photosynthesis via stomatal response to environmental conditions must also be characterized in production models. This has been implicit in ecophysiological modeling approaches through terms for vapor pressure deficit and foliage water potentials, but has also been incorporated into remote sensing models (e.g., Equation 3).

The evidence for resource-use optimization and a convergence of production efficiency on a mass per unit area basis suggests that "ecological integrators" can be used to simplify the biophysical and physiological components of models used to estimate net annual carbon fluxes over large areas. Simplifications that are implied by the evaluation of functional convergence provided here enable remote sensing to provide a robust approach to NPP evaluation. Moreover, these simplifications can be incorporated into models that utilize remotely sensed data, provided that the coupling between the vegetation and atmosphere is adequately characterized.

Summary of Functional Convergence Analyses

A great deal of evidence reported in recent years suggests that natural selection of traits that confer fitness results in suites of traits that are closely related (including, for example, c, leaf longevity and relative growth rate). Several traits have been identified as "ecological integrators" that characterize a given growth-form and its associated allocation of resources. These traits reflect trade-offs in costs versus benefits of resource acquisition, allocation and utilization. Resource shortages of any sort will lead to adjustments of light capture. Hence light capture serves as an integrator of resource status. SVI s, which are linked to foliage display and light absorption by vegetation, are thus linked with many other plant functions.

A decoupling of the canopy light environment from the physical climate system through stomatal control may, however, also decouple the links between foliage display, light absorption and reflectance properties detected with satellite sensors, thus the NPP of terrestrial ecosystems. Indicators of the short-term links between the physical climate system and canopy light absorption must therefore be incorporated into production models driven with remotely sensed data, particularly in forest ecosystems where the canopy is more tightly linked with the atmosphere and less to the net radiation environment.

There is ample evidence that allocation trade-offs associated with responses to combined multiple stresses results in an optimization of resource-use through selection, but this optimization is evident only on a leaf mass basis, not on a leaf area or volume basis. There is also strong evidence that a maximization of fitness can be equated with a maximization of carbon gain. The allocation of assimilated carbon, however, varies by growth strategy such that differences in respiratory costs relative to carbon (energy) gain do not result in a convergence on a narrow range of light-use efficiencies (or dry matter yields of energy) among plant species. For example, proportionately more resources are allocated to non-photosynthetic components in plant functional types adapted to resource-poor environments, which in turn results in greater respiratory costs relative to carbon gain. Thus decoupling of light harvesting and utilization may also occur via biochemical channels. These factors affect the linkages between maximized carbon assimilation, PAR utilization, and production models driven by light harvesting.

Thus, evaluation of the available evidence, including the results of Chapters II and III, force a rejection of the functional convergence hypothesis with respect to convergence in a narrow range of n. More importantly, however, the analyses suggest that functional convergence results in a narrow range in g, both among and between plant functional types, and in the R:A ratio within functional types. Thus, the hypothesis is valid when interpreted in a more restricted sense than originally proposed. These conclusions are based on the results of the production efficiency modeling (Chapter II) and the carbon flux modeling (Chapter III), which clearly showed variability in n that was related to the respiratory costs associated with the different life histories of quaking aspen versus black spruce. The conclusions were also strongly supported by an assessment of the hypothesis in terms of evolutionary ecology; i.e., the inter-disciplinary research on resource-based growth constraints (stresses), associated resource trade-offs, the costs versus benefits of various allocation strategies (particularly growth versus defense allocation), the evidence for optimization of resource-use efficiency, and the associated evidence for maximization of carbon gain and fitness. The conclusions should be tested further with additional measurements of light absorption and whole-ecosystem production, as such data become available.

Synopsis and Conclusions

Remote sensing provides observational data sets that can be used in modeling and monitoring net primary production of the terrestrial biosphere. Improvements in NPP modeling may help to reduce uncertainty in global carbon budgets, thus in simulations that incorporate vegetation feedbacks on climate. The task of accurately estimating NPP globally with remotely sensed data can be simplified if light-use efficiency varies little or at least predictably in different vegetation types. The research reported herein has shown that factors related to light utilization are at least as relevant as light harvesting, and that both are affected by evolutionary tuning of plant attributes.

There are different ways in which light harvesting and light utilization may become uncoupled. The most obvious and best documented way is via physical factors such as temperature that can have a constraining effect on the rate of photosynthetic reactions. Also, light may be absorbed by foliage but remain unavailable for utilization owing to resource limitations (typically water) that induce various degrees of stomatal control by the plant. Stomatal control is also linked to coupling of the plant with the atmosphere in the boundary layer between them, and the depth of this boundary layer varies with the aerodynamic roughness of a plant canopy and other physical and biotic factors related to surface energy balance.

The second manner in which light harvesting and light utilization may become uncoupled was examined in some detail as part of this research. Whereas the amount of energy harvested by a plant, in the form of photosynthetically active radiation, may scale with the amount of resources available to a plant and thus its foliage display and carbon gain, allocation of assimilate for different processes may quite often be for non-growth purposes, including defense, structural support and reproduction. The various adaptations in plants reflect these different utilization and allocation patterns in the energy costs required for construction and maintenance of various tissues, and these costs are in turn reflected in the life span of the plant and its foliage (i.e., the "payback period").

The differential in energy costs associated with functional adaptations is reflected in both the biomass density of various plant tissues and the ratio of CO_2 respiration (R) to assimilation (A); referred to as the R:A ratio or "respiration efficiency." The link between respiratory costs and biomass was evident in an analysis of resource use optimization (summarized at the end of Chapter IV), which converged for a range of different functional types (plants with a suite of related life history traits) on a mass per unit area basis. The mass basis for convergence in resource use efficiency reflects the amount of energy (carbon) required for various plant

100

processes better than area-based indices. Thus R:A ratio and biomass density are indicators of the amount of energy utilized for net primary production.

It has been suggested that a narrow range in PAR utilization (ε_n) has emerged through selection of traits that enhance the efficiency of resource-use. The results of model simulations conducted in Chapter III suggest, however, that convergence has been on the basis of a maximized carbon assimilation rather than net production. In other words, there is a convergence on the amount of CO_2 assimilated by a canopy per unit APAR (ε_g) (see Figure 23), rather than on the amount of NPP per unit APAR (ε_n) (Figure 17). The results of an analysis of the amount of PAR harvested by boreal forest stands, as estimated with a simple production efficiency model driven with remotely sensed observations, support these observations, as do most other studies of NPP relative to PAR harvesting (see Chapter II summary). Variability in ε_n increased as the number of plant functional types increased (Figure 8).

Simulated R:A ratios derived with an ecophysiological model for a range of measured stand variables showed that much of the variability in ε_n was explained by respiratory carbon costs relative to carbon assimilation (see Figure 19). Moreover, the R:A ratio was less variable within than between functional types owing to the different carbon allocation costs. The model simulations also showed, however, that the relationship between the R:A ratio and above-ground biomass was largely invariant among spruce stands (a slow-growing, long-lived species), and was affected by stress in some aspen stands (a fast-growing, short-lived species). This observation was related to the covariation of respiration with assimilation, as both increased with respiring biomass amount (sapwood and foliage). Thus, whereas the R:A ratio was a good predictor of ε_n, above-ground biomass was not a reliable predictor of the R:A ratio. Additional research is needed to explore this issue further in a wider range of plant functional types.

Approaches to NPP modeling with remote sensing (i.e., production efficiency models), which up to now have been based on light harvesting rather than utilization, require augmentation with variables related to utilization (e.g., respiration costs reflected in plant ecophysiology). In particular, production efficiency models require variables that account for differences in the amount of energy harvested relative to the amount of energy expended in the production of a given amount of dry matter. Augmenting production efficiency models for the effect of variability in PAR utilization requires: (i) estimating the amount of respiring biomass rather than total existing biomass, as the former alone is actively respiring, (ii) accounting for stresses, which modify not only ε_n but also the R:A ratio and ε_g, (iii) accounting for categorical differences in the R:A ratio between functional types. The first requirement is particularly relevant to woody vegetation as much of the existing biomass may be for structural support or defense, rather than growth. The second requirement has long been utilized in ecophysiological models, but observations summarized at the end of Chapter IV suggest such models may be functionally simpler than their design implies. Similarly, terms that account for stress can be incorporated in the functionally simpler but spatially explicit production efficiency models (e.g., Equation 3). Recent work accounting for stresses in a production efficiency model also support this conclusion (Runyon et al. 1994; Goward et al. 1994). The third requirement may account for much of observed variability in n alone, as the simulation modeling results showed (see Chapter III summary). This is the first known study to specifically document the effect of respiratory carbon costs on variability in n. This conclusion is particularly relevant to biomes in which a number of plants of various functional types, with associated differences in R:A ratios, are present (e.g., boreal, temperate and tropical forests). Variability in n in other biomes (e.g., semi-arid woodlands, grasslands) may be more closely linked to stresses associated with stomatal control than with respiratory costs, and yet other biomes may be intermediate between the two.

In conclusion, the research reported herein demonstrates that modifications to some of the assumptions of production efficiency models are required. In particular, the hypothesis of functional convergence, which suggests that n should occupy a narrow range of values among functional types, was shown to be more applicable to g between functional types, and the R:A ratio within functional types. Similarly, a suggestion that respiration may be a nearly constant proportion of assimilation through time was shown to be untenable when extended to a wider range of plant functional types. The results of the experimental work (Chapters II and III) and the integration and interpretation of evidence from evolutionary ecology (Chapter IV) specifically demonstrate that modifications to production efficiency models should include; (i) the use of APAR as a relatively invariant driver of gross production, (ii) the R:A ratio as a reliable determinant of variability in n, (iii) moisture stress regulation through stomatal control as a modifier of g and R:A. Additional analyses should be conducted with coincident experimental data collected under a wide range of conditions, as such data become more generally available. Further advancement of techniques that allow realistic simplification of NPP models will have potentially significant implications for the monitoring of terrestrial carbon dynamics at the global scale.

Appendix: Glossary of Acronyms, Definitions, Terms, and Symbols

Terms related to physiological ecology and production biology

AANPP	Annual above-ground net primary production (NPP) (reported as g m^{-2} yr^{-1}).
Assimilation	The uptake of carbon dioxide (CO_2) by plants through photosynthesis, expressed herein as the mass of CO_2 (g) assimilated per unit area (m) per unit time (day).
GDD	Growing Degree-Days, a thermal index derived from the sum of air temperatures above a specified threshold (typically 5° C).
GPP	Gross Primary Production, equivalent to NPP - R$_a$
LAI	Leaf Area Index, an unitless index of the amount of stacked leaf area (m^2). relative to the surface area beneath (m^2). The values used herein are reported as one-sided or projected values.
LMA	Leaf Mass per unit Area (e.g., g m^{-2}).
LWR	Leaf Weight Ratio, the fraction of total plant biomass allocated to foliage.
NEP	Net Ecosystem Production, equivalent to NPP - R$_h$.
NPP	Net Primary Production, total biomass production through time, typically expressed in terms of production per unit area per unit time (e.g., g m^{-2} yr^{-1}) and equivalent to the difference between total assimilation and respiration.
R$_a$	Autotrophic respiration, the release of CO_2 by plants through the construction of new biomass (growth respiration) and the support of existing biomass (maintenance respiration). Expressed herein in the same units as assimilation.
R:A	The ratio of maintenance respiration to assimilation, also referred to herein as respiration efficiency (unitless).
R$_h$	Heterotrophic respiration, CO_2 released from soils through microbial metabolism of organic (i.e., carbon based) compounds.
RGR	Relative Growth Rate, the increment of dry weight of a single plant per unit weight per unit time, equivalent to the product of SLA, LWR and the average unit leaf rate of assimilation.
SLA	Specific Leaf Area, the area of foliage relative to it mass (e.g., m^2 g^{-1}).
SDD	Soil Degree-Days, a thermal index derived from the sum of soil temperatures above a specified threshold (typically 0° C).
Y$_m$	The proportion of assimilated carbon not lost to maintenance respiration (synonymous with the quantity 1-R:A).
Y$_g$	The proportion of assimilated carbon not lost to growth respiration.
	Epsilon, the dry matter yield of harvested PAR, alterately referred to as light use efficiency, radiation use efficiency, and the dry matter yield of energy. Referred to herein more generally as PAR utilization. Units are in terms of biomass production per unit energy (e.g., g MJ^{-1}).
a	in terms of the utilization of absorbed PAR.
i	in terms of the utilization of intercepted PAR.
n	expressed explicitly in terms of net production, synonymous with a or i.
g	in terms of gross production, or total CO_2 assimilation (g CO_2 MJ^{-1}).
g^*	The potential or unstressed form of g. Related to, but not synonymous with . The quantum yield of photosynthesis, or assimilation per unit energy, inferred from the initial slope of a photosynthesis light-response curve. It generally increases with decreasing temperature, above freezing, in C3 plants.

c   Photosynthetic capacity, the maximum (potential) rate of photosynthetic carbon gain per unit leaf area per unit time (e.g., mol m^{-2} s^{-1}).

<u>Terms related to environmental physics</u>

Albedo	The fraction (or percent) of incident radiation reflected by a surface. Albedo can be specified, for example, as spectral (specific to a wavelength range) or hemispheric (integrated over all wavelengths and view directions).
APAR	Absorbed Photosynthetically Active Radiation, an index of PAR harvesting by vegetation, used here in terms of a daily or an annual amount (MJ). See also APAR.
G_i	Kauth-Thomas Greenness, a spectral vegetation index (SVI).
IPAR	Intercepted Photosynthetically Active Radiation, an index of PAR harvesting by vegetation, used here in terms of a daily or an annual amount (MJ). See also IPAR. Note that absorbed PAR differs from intercepted PAR in that the former includes PAR albedo reflected from the background substrate into the overlying canopy.
APAR	The annual sum of absorbed PAR (MJ). See also APAR.
IPAR	The annual sum of intercepted PAR (MJ). See also IPAR.
MJ	Megajoule, a million joules, where a joule is a unit of energy equal to the work done when a force of 1 newton acts through a distance of 1 meter (one joule per second is equal to one watt).
NDVI	The Normalized Difference Vegetation Index, a widely used unitless spectral vegetation index (SVI) that utilizes the ratio of the difference between the reflectances of near-infrared and visible wavelength bands to the sum of the same, i.e., [IR-Vis]/[IR+Vis].
PAR	Photosynthetically Active Radiation (synonymous with Incident PAR).
PAR	Same as PAR (incident).
f_{par}	The fraction of incident PAR intercepted or absorbed by the vegetation canopy.
f_{apar}	The fraction of incident PAR absorbed by the vegetation canopy.
f_{ipar}	The fraction of incident PAR intercepted by the vegetation canopy.
SVI	Spectral Vegetation Index, a mathematical combination of surface spectral reflectance in different wavelength bands that enhance the signature of vegetation relative to other scene components (such as soil background).

References

Abrams, M. D., M. E. Kubiske and S. A. Mostoller. 1994. Relating wet and dry year ecophysiology to leaf structure in contrasting temperate tree species, *Ecology* 75 (1): 123-133.

Agust , S., S. Enr quez, H. Frost-Christensen, K. Sand-Jensen and C. M. Duarte. 1994. Light harvesting among photosynthetic organisms, *Functional Ecology* 8 : 273-279.

Ahlgren, C. E. 1957. Phenological observations of nineteen native tree species in northeastern Minnesota, *Ecology* 38 (4): 622-628.

Alban, D. H., D. A. Perala and B. E. Schlaegel. 1978. Biomass and nutrient distribution in aspen, pine, and spruce stands on the same soil type in Minnesota, *Canadian Journal of Forest Research* 8 : 290-299.

Allen, J. C. 1976. A modified sine wave method for calculating degree days, *Environmental Entomology* 5 (3): 388-396.

Amthor, J. S. 1995. Higher plant respiration and its relationships to photosynthesis. Pages 71-101 *in* E.-D. Schulze and M. M. Caldwell, Ed., *Ecophysiology of Photosynthesis.* Berlin, Springer Verlag.

Anekonda, T. S., R. S. Criddle, W. J. Libby, R. W. Breidenbach and L. D. Hansen. 1994. Respiration rates predict differences in growth of coast redwood, *Plant, Cell and Environment* 17 : 197-203.

Asrar, G., M. Fuchs, E. T. Kanemasu and J. L. Hatfield. 1984. Estimating absorbed photosynthetic radiation and leaf area index from spectral reflectance in wheat, *Agronomy Journal* 76 : 300-306.

Asrar, G., E. T. Kanemasu, R. D. Jackson and P. J. Pinter. 1985. Estimation of total above ground phytomass production using remotely sensed data, *Remote Sensing of Environment* 17 : 211-220.

Asrar, G., E. T. Kanemasu, G. P. Miller and R. L. Weiser. 1986. Light interception and leaf area estimates from measurements of grass canopy reflectance, *IEEE Transactions on Geoscience and Remote Sensing* GE24 : 76-82.

Atjay, G. L., P. Ketner and Duvigneaud. 1979. Terrestrial primary production and phytomass. Pages *in* B. Bolin, Ed., *The Global Carbon Cycle.* New York, John Wiley & Sons.

Bacastow, R. B., C. D. Keeling and T. P. Whorf. 1985. Seasonal amplitude increase in atmospheric CO_2 concentration at Mauna Loa, Hawaii, 1959-1982, *Journal of Geophysical Research* 90 : 10529-10540.

Badhwar, G. D. 1984a. Classification of corn and soybeans using multitemporal Thematic Mapper data, *Remote Sensing of Environment* 16 : 175-182.

Badhwar, G. D. 1984b. Use of Landsat-derived profile features for spring small-grains classification, *International Journal of Remote Sensing* 5 (5): 783-797.

Badhwar, G. D., R. B. Macdonald, F. G. Hall and J. G. Carnes. 1986. Spectral characterization of biophysical characteristics in a Boreal Forest: Relationship between Thematic Mapper band reflectance and leaf area index for aspen, *IEEE Transactions on Geoscience and Remote Sensing* GE-24 (3): 322-326.

Baldocchi, D. D., S. B. Verma and D. E. Anderson. 1987. Canopy photosynthesis and water-use efficiency in a deciduous forest, *Journal of Applied Ecology* 24 : 251-260.

Ball, J. T., I. E. Woodrow and J. A. Berry. 1987. A model predicting stomatal conductance and its contribution to the control of photosynthesis under different environmental conditions, In: *Progress in Photosynthesis Research: Proceedings of the 7th International Congress on Photosynthesis*, pp. 221-224, Providence, Rhode Island.

Bazilevich, N. I., L. Y. Rodin and N. N. Rozov. 1971. Geographical aspects of biological productivity, *Soviet Geography* 12 : 293-317.

Bazzaz, F. A., J. S. Coleman and S. R. Morse. 1990. Growth responses of seven major co-occuring tree species of the northeastern United States to elevated CO_2, *Canadian Journal of Forest Research* 20 : 1479-1484.

Begon, M., J. L. Harper and C. R. Townsend. 1990. Life-history variations. Pages 473-509 *in* M. Begon, J. L. Harper and C. R. Townsend, Ed., *Ecology*. Oxford, Blackwell Scientific Publications.

B gu , A. 1991. Modeling hemispherical and directional radiative fluxes in regular-clumped canopies, *Remote Sensing of Environment* 40 (3): 219-230.

Biscoe, P. V., R. K. Scott and J. L. Monteith. 1975. Barley and its environment III. Carbon budget of the stand, *Journal of Applied Ecology* 12 : 269-293.

Bj rkman, O. 1981. Responses to different quantum flux densities. Pages 57-108 *in* O. L. Lange, P. S. Nobel, C. B. Osmond and H. Ziegler, Ed., *Physiological Plant Ecology*. Berlin, Springer-Verlag.

Bloom, A., F. S. Chapin III and H. Mooney. 1985. Resource limitation in plants - an economic analogy, *Annual Review Ecology and Systematics* 16 : 363-392.

Bonan, G. B. 1990. Carbon and nitrogen cycling in North American boreal forests. II. Biogeographic patterns., *Canadian Journal of Forest Research* 20 : 1077-1088.

Bonan, G. B. 1991a. Atmosphere-biosphere exchange of carbon dioxide in boreal forests, *Journal of Geophysical Research* 96 (D4): 7301-7312.

Bonan, G. B. 1991b. A biophysical surface energy budget analysis of soil temperature in the boreal forests of interior Alaska, *Water Resources Research* 27 (5): 767-781.

Bonan, G. B. 1993a. Importance of leaf area index and forest type when estimating photosynthesis in boreal forests, *Remote Sensing of Environment* 43 : 303-314.

Bonan, G. B. 1993b. Physiological controls of the carbon balance of boreal forest ecosystems, *Canadian Journal of Forest Research* 23 (7): 1453-1471.

Botkin, D. B., J. F. Janek and J. R. Wallis. 1972. Some ecological consequences of a computer model of forest growth, *Journal of Ecology* 60 : 849-872.

Box, E. O. 1978. Geographical dimensions of terrestrial net and gross primary productivity, *Radiation and Environmental Biophysics* 15 : 305-322.

Briggs, G. M., T. W. Jurik and D. M. Gates. 1986. A comparison of rates of aboveground growth and carbon dioxide assimilation by aspen on sites of high and low quality, *Tree Physiology* 2 : 29-34.

Brown, S. and A. E. Lugo. 1982. The storage and production of organic matter in tropical forests and their role in the global carbon cycle, *Biotropica* 14 (3): 161-187.

Brown, S. and A. E. Lugo. 1984. Biomass of tropical forests: A new estimate based on forest volumes, *Science* 223 : 1290-1293.

Bryant, J. P., F. S. Chapin and D. R. Klein. 1983. Carbon/nutrient balance of boreal plants in relation to vertebrate herbivory, *Oikos* 40 : 357-368.

Burns, T. P. 1992. Adaptedness, evolution and a hierarchical concept of fitness, *Journal of Theoretical Biology* 154 : 219-237.

Cannell, M. G. R., R. Milne, L. J. Sheppard and M. H. Unsworth. 1987. Radiation interception and productivity in willow, *Journal of Applied Ecology* 24 : 261-278.

Caswell, H. 1982. Life history theory and the equilibrium status of populations, *American Naturalist* 120 : 317-339.

Chabot, B. F. and D. J. Hicks. 1982. The ecology of leaf life-spans, *Annual Review of Ecology and Systematics* 13 : 229-259.

Chapin, F. S., A. J. Bloom, C. B. Field and R. H. Waring. 1987. Plant responses to multiple environmental factors, *BioScience* 37 : 49-57.

Chapin, F. S. 1989. The cost of tundra plant structures: evaluation of concepts and currencies, *American Naturalist* 133 : 1-19.

Chapin, F. S. 1993. Functional role of growth forms in ecosystem and global processes. Pages 287-312 *in* J. R. Ehleringer and C. B. Field, Ed., *Scaling Physiological Processes*. San Diego California, Academic Press, Inc.

Chapin, F. S., E.-D. Schulze and H. A. Mooney. 1990. The ecology and economics of storage in plants, *Annual Review of Ecology and Systematics* 21 : 423-447.

Chen, J.-L., J. F. Reynolds, P. C. Harley and J. D. Tenhunen. 1993. Coordination theory of leaf nitrogen distribution in a canopy, *Oecologia* 93 : 63-69.

Choudhury, B. J. 1987. Relationships between vegetation indices, radiation absorption, and net photosynthesis evaluated by a sensitivity analysis, *Remote Sensing of Environment* 22 : 209-233.

Ciais, P., P. P. Tans, M. Trolier, J. W. C. White and R. J. Francey. 1995. A large northern hemisphere terrestrial CO2 sink indicated by the $^{13}C/^{12}C$ ratio of atmospheric CO_2, *Science* 269 (5227): 1098-1101.

Coley, P. D. 1986. Costs and benefits of defense by tannins in a neotropical tree, *Oecologia* 70 : 238-241.

Coley, P. D., J. P. Bryant and F. S. Chapin. 1985. Resource availability and plant anti-herbivore defense, *Science* 230 : 895-899.

Collatz, G. J., J. T. Ball, C. Grivet and J. A. Berry. 1991. Physiological and environmental regulation of stomal conductance, photosynthesis and transpiration: a model that includes a laminar boundary layer, *Agricultural and Forest Meteorology* 54 : 107-136.

Conway, T. J., P. P. Tans, L. S. Waterman, K. W. Thoning, D. R. Kitzis, K. A. Masarie and N. Zhang. 1994. Evidence for interannual variability of the carbon cycle from the National Oceanic and Atmospheric Administration / Climate Monitoring and Diagnostics Laboratory global air sampling network, *Journal of Geophysical Research* 99 : 22831-22855.

Cowan, J. R. 1986. Economics of carbon fixation in higher plants. Pages *in* T. J. Givnish, Ed., *On the economics of plant form and function.* Cambridge, Cambridge University Press.

Crist, E. P. 1985. A TM tasseled cap equivalent transformation for reflectance factor data, *Remote Sensing of Environment* 17 : 301-306.

D Arrigo, R., G. C. Jacoby and I. Y. Fung. 1987. Boreal forests and atmosphere - biosphere exchange of carbon dioxide, *Nature* 329 (24): 321-323.

Dang, Q., H. Margolis, M. R. Coyea, M. Sy, G. J. Collatz and D. Yue. In press. Environmental controls on photosynthesis and stomatal conductance of boreal forest tree species, *Tree Physiology.*

Daughtry, C. S. T., K. P. Gallo, S. N. Goward, S. D. Prince and W. P. Kustas. 1992. Spectral estimates of absorbed radiation and phytomass production in corn and soybean canopies, *Remote Sensing of Environment* 39 : 141-152.

Dennett, D. C. 1995. *Darwin s Dangerous Idea*, Simon and Schuster, New York.

Denning, A. S., I. Y. Fung and D. Randall. 1995. Latitudinal gradient of atmospheric CO_2 due to seasonal exchange with land biota, *Nature* 376 : 240-243.

Desjardins, R. L., J. L. MacPherson, P. Alvo and P. H. Schuepp. 1985. Measurements of turbulent heat and CO_2 exchange over forests from aircraft. Pages 645-658 *in* B. A. Hutchinson and B. B. Hicks, Ed., *Forest-Atmosphere Interactions.* Hingman Massachusetts, Reidel Press.

Detwiler, R. P. and C. A. S. Hall. 1988. Tropical forests and the global carbon cycle, *Science* 239 : 42-47.

Dickinson, R. E. and A. Henderson-Sellers. 1988. Modelling tropical deforestation: A study of GCM land-surface parameterizations, *Quarterly Journal of the Royal Meteorological Society* 114 : 439-462

Dickinson, R. E. 1995. Land processes in climate models, *Remote Sensing of Environment* 51 : 27-38.

Dickinson, R. E., A. Henderson-Sellers, P. J. Kennedy and M. F. Wilson. 1986. *Biosphere-atmosphere transfer scheme (BATS) for the NCAR community climate model*, National Center for Atmospheric Research, Technical Note TN-275+STR.

Donovan, L. A. and J. R. Ehleringer. 1994. Carbon isotope discrimination, water-use efficiency, growth, and mortality in a natural shrub population, *Oecologia* 100 (3): 347-354.

Doust, J. L. 1989. Plant reproductive strategies and resource allocation, *Trends in Ecology and Evolution* 4 (8): 230-234.

Dye, D. 1993. Satellite estimation of the global distribution and interannual variability of photosynthetically active radiation. University of Maryland, Doctoral Thesis

Ellsworth, D. S. and P. B. Reich. 1993. Canopy structure and vertical patterns of photosynthesis and related leaf traits in a deciduous forest, *Oecologia* 96 : 169-178.

Emanuel, W. R., H. H. Shugart and M. P. Stevenson. 1985. Climate change and the broad-scale distribution of terrestrial ecosystem complexes, *Climate Change* 7 : 29-43.

Erhleringer, J. H. and K. S. Werk. 1986. Modifications of solar radiation absorption in patterns and implications for carbon gain at the leaf level. Pages 57-82 *in* T. J. Givnish, Ed., *On the Economics of Plant Form and Function*. Cambridge, Cambridge University Press.

Ehleringer, J. R. 1993. Variation in leaf carbon isotope discrimination in *Encilia farinosa*: implications for growth, competition and drought survival, *Oecologia* 95 (3): 340-346.

Fagerstr m, T., S. Larsson and O. Tenow. 1987. On optimal defence in plants, *Functional Ecology* 1 : 73-81.

Fan, S.-M., S. C. Wofsy, P. S. Bakwin, D. J. Jacob and D. R. Fitzjarrald. 1990. Atmosphere-biosphere exchange of CO_2 and O_3 in the central Amazon forest, *Journal of Geophysical Research* 95 (D10): 16851-16864.

Farquhar. 1989. Models of integrated photosynthesis of cells and leaves, *Philosophical Transactions of the Royal Society of London (B)* 323 : 357-367.

Field, C. B. 1983. Allocating leaf nitrogen for the maximization of carbon gain: leaf age as a control on the allocation program, *Oecologia* 56 : 341-347.

Field, C. B. 1991. Ecological scaling of carbon gain to stress and resource availability. Pages 35-65 *in* H. A. Mooney, W. E. Winner and E. J. Pell, Ed., *Response of plants to multiple stresses*. San Diego, Academic Press.

Field, C. B. and H. A. Mooney. 1986. The photosynthesis - nitrogen relationship in wild plants. Pages 681-698 *in* T. J. Givnish, Ed., *On the Economy of Plant Form and Function*. Cambridge, England, Cambridge University Press.

Field, C. B., J. T. Randerson and C. M. Malmstr m. 1995. Global net primary production: combining ecology and remote sensing, *Remote Sensing of Environment* 51 (1): 74-88.

Forrest, S. 1993. Genetic algorithms - principles of natural selection applied to computation, *Science* 261 (5123): 872-878.

Forseth, I. N. and J. R. Ehleringer. 1982. Ecophysiology of two solar tracking desert annuals 2. leaf movements, water relations and microclimate, *Oecologia* 54 (1): 41-49.

Friend, A. D., H. H. Shugart and S. W. Running. 1993. A physiology-based gap model of forest dynamics, *Ecology* 74 (3): 792-797.

Gadgil, M. and W. Bossert. 1970. Life history consequences of natural selection, *American Naturalist* 104 : 1-24.

Geiger, D. R. and J. C. Servaites. 1991. Carbon allocation and response to stress. Pages 103-127 in H. A. Mooney, W. E. Winner and E. J. Pell, Ed., *Response of Plants to Multiple Stress*. San Diego, Academic Press.

Gholtz, H. L., S. A. Vogel, W. P. Cropper, K. McKelvey, K. C. Ewel, R. O. Tesky and P. J. Curran. 1991. Dynamics of canopy structure and light interception in *Pinus elliottii* stands, north Florida, *Ecological Monographs* 61 (1): 33-51.

Givnish, T. J. 1982. On the adaptive significance of leaf height in forest herbs, *American Naturalist* 120 (3): 353-381.

Gleeson, S. K. 1993. Optimization of tissue nitrogen and root-shoot allocation, *Annals of Botany* 71 : 23-31.

Gleeson, S. K. and D. Tilman. 1994. Plant allocation, growth rate and successional status, *Functional Ecology* 8 : 543-550.

Goel, N. and W. Qin. 1994. Influence of canopy architecture on various vegetation indices and LAI and Fpar: simulation model results, *Remote Sensing Reviews* 10 : 309-347.

Gould, S. J. and R. C. Lewontin. 1979. The spandrels of San Marco and the Panglossian paradigm: A critique of the the adaptationist programme, *Proceedings of the Royal Society of London (B)* 205 : 581-598.

Goward, S. N. and K. F. Huemmrich. 1992. Vegetation canopy PAR absorbance and the normalized difference vegetation index: an assessment using the SAIL model, *Remote Sensing of Environment* 39 : 119-140.

Goward, S. N. and S. D. Prince. 1995. Transient effects of climate on vegetation dynamics: satellite observations, *Journal of Biogeography* 22 : 549-563.

Goward, S. N., C. J. Tucker and D. G. Dye. 1985. North American vegetation patterns observed with the NOAA-7 advanced very high resolution radiometer, *Vegetatio* 64 : 3-14.

Goward, S. N., R. H. Waring and D. G. Dye. 1994. Ecological remote sensing at OTTER: Macroscale satellite observations, *Ecological Applications* 4 (2): 322-343.

Grace, J., J. Lloyd, J. McIntyre, A. C. Miranda, P. Meir, H. S. Miranda, C. Nobre, J. Moncrieff, J. Massheder, Y. Malhi, I. Wright and J. Gash. 1995. Carbon dioxide uptake by an undisturbed tropical rain forest in southwest Amazonia, 1992 to 1993, *Science* 270 : 778-780.

Griffin, K. L. 1994. Calorimetric estimates of construction cost and their use in ecological studies, *Functional Ecology* 8 : 551-562.

Grigal, D. F. and H. F. Arneman. 1970. Quantitative relationships among vegetation and soil classifications from northeastern Minnesota, *Canadian Journal of Botany* 48 : 555-566.

Grime, J. P. 1977. Evidence for the existence of three primary strategies in plants and its relevance to ecological and evolutionary theory, *American Naturalist* 111 : 1169-1194.

Grime, J. P. 1979. *Plant Strategies and Vegetation Processes*, John Wiley and Sons, London.

Gutschick, V. P. 1981. Evolved strategies in nitrogen acquisition by plants, *American Naturalist* 118 : 607-637.

Gutschick, V. P. 1993. Nutrient-limited growth rates: Roles of nutrient-use efficiency and of adaptations to increase uptake rate, *Journal of Experimental Botany* 44 (258): 41-51.

Hall, F. G., D. B. Botkin, D. E. Strebel, K. D. Woods and S. J. Goetz. 1991a. Large scale patterns of forest succession as determined by remote sensing, *Ecology* 72 (2): 628-640.

Hall, F. G., K. F. Heummrich, D. E. Strebel, S. J. Goetz, J. E. Nickeson and K. D. Woods. 1992a. *Biophysical, morphological, canopy optical property, and productivity data from the Superior National Forest*, NASA, Technical Memorandum 104568 (available from the U. S. National Technical Information Service).

Hall, F. G., K. F. Huemmrich, S. J. Goetz, P. J. Sellers and J. E. Nickeson. 1992b. Satellite remote sensing of surface energy balance: Successes, failures and issues in FIFE, *Journal of Geophysical Research* 97 (D17): 19061-19089.

Hall, F. G., Y. Shimabukuro and K. F. Huemmrich. 1995. Remote sensing of forest biophysical structure in boreal stands of *Picea mariana* using mixture decomposition and geometric reflectance models, *Ecological Applications* 5 (4): 993-1013.

Hall, F. G., D. E. Strebel, J. E. Nickeson and S. J. Goetz. 1991b. Radiometric rectification: Toward a common radiometric response among multi-date, multi-sensor images, *Remote Sensing of Environment* 35 : 11-27.

Hansen, J., D. Johnson, A. Lacis, S. Lebedeff, P. Lee, D. Rind and G. Russell. 1981. Climate impact of increasing atmospheric carbon dioxide, *Science* 213 (4511): 957-966.

Harden, J. W., E. T. Sunquist, R. F. Stallard and R. K. Mark. 1992. Dynamics of soil carbon during deglaciation of the Laurentide ice sheet, *Science* 258 : 1921-1923.

Harper, J. L. 1989. The value of a leaf, *Oecologia* 80 : 53-58.

Heinselman, M. L. 1973. Fire in the virgin forests of the Boundary Waters Canoe Area, Minnesota, *Quaternary Research* 3 : 329-382.

Hilbert, D. W. 1990. Optimization of plant root:shoot ratios and internal nitrogen concentration, *Annals of Botany* 66 : 91-99.

Hirose, T. 1987. A vegetative plant growth model: adaptive significance of phenotypic plasticity in matter partitioning, *Functional Ecology* 1 : 195-202.

Hollinger, D. Y. 1989. Canopy organization and foliage photosynthetic capacity in a broad-leaved evergreen montane forest, *Functional Ecology* 3 : 53-62.

Hom, J. L. and W. C. Oechel. 1983. The photosynthetic capacity, nutrient content, and nutrient use efficiency of different needle age-classes of black spruce (*Picea mariana*) found in interior Alaska, *Canadian Journal of Forest Research* 13 : 834-839.

Horn, H. 1971. *The Adaptive Geometry of Trees*, Princeton University Press, Princeton, NJ.

Houghton, R. A. 1987. Terrestrial metabolism and atmospheric CO_2 concentrations, *BioScience* 37 (9): 672-678.

Houghton, R. A., J. E. Hobbie, J. M. Mellilo, B. M. III, B. J. Peterson, G. R. Shaver and G. M. Woodwell. 1983. Changes in the carbon content of terrestrial biota and soils between 1860 and 1980: a net release of CO_2 to the atmosphere, *Ecological Monographs* 53 : 235-262.

Huemmrich, K. F. 1995. An analysis of remote sensing of absorbed photosynthetically active radiation in forest canopies. University of Maryland, College Park, Doctoral Thesis.

Hunt, E. R. and S. Running. 1992a. Effects of climate and lifeform on dry matter yield ( ) from simulations using Biome-BGC, In: *International Geoscience and Remote Sensing Symposium*, pp. 1631-1633, Clear Lake, Texas.

Hunt, E. R. and S. W. Running. 1992b. Simulated dry matter yields for aspen and spruce stands in the North American boreal forest, *Canadian Journal of Remote Sensing* 18 (3): 126-133.

Hunt, E. R. 1994. Relationship between woody biomass and PAR conversion efficiency for estimating net primary production from NDVI, *International Journal of Remote Sensing* 15 : 1725-1730.

Hunt, R. and P. S. Lloyd. 1987. Growth and partitioning, *New Phytologist* 106 : 235-249.

Hutchinson, B. A., D. R. Matt, R. T. McMillen, L. J. Gross, S. J. Tajchman and J. M. Norman. 1986. The architecture of a deciduous forest canopy in eastern Tennessee, USA, *Journal of Ecology* 74 : 635-646.

Iacobelli, A. and J. H. McCaughey. 1993. Stomatal conductance in a northern temperate deciduous forest - temporal and spatial patterns, *Canadian Journal of Forest Research* 23 (2): 245-252.

Jacoby, G. and R. D Arrigo. 1995. Tree ring width and density evidence of climatic and potential forest change in Alaska, *Global Biogeochemical Cycles* 9 (2): 227-234.

Jarvis, P. G. 1976. The interpretation of variations in the leaf water potential and stomatal conductance found in canopies in the field, *Philosophical Transactions of the Royal Society of London (B)*, 273 : 593-610.

Jarvis, P. G. and J. W. Leverentz. 1983. Productivity of temperate, deciduous and evergreen forests. Pages 233-280 *in* O. L. Lange, P. S. Nobel, C. B. Osmond and H. Ziegler, Ed., *Physiological Plant Ecology IV. Ecosystem Processes: Mineral Cycling, Productivity and Man s Influence*. New York, Springer-Verlag.

Jordan, C. F. 1971. Productivity of a tropical forest and its relation to a world pattern of energy storage, *Journal of Ecology* 59 : 127-142.

Kachi, N. and I. H. Rorison. 1989. Optimal partitioning between root and shoot in plants with contrasted growth rates in response to nitrogen availability and temperature, *Functional Ecology* 3 : 549-559.

Kauth, R. J., P. F. Lambeck, W. Richarson, G. S. Thomas and A. P. Pentland. 1979. Feature extraction applied to agricultural crops as seen by LANDSAT, In: *LACIE Symposium*, NASA, pp. 705-721, Johnson Space Center, Houston Texas.

Kauth, R. J. and G. S. Thomas. 1976. The tassel-cap: a graphic description of the spectral-temporal development of agricultural crops as seen by LANDSAT, In: *Proceeding of Machine Processing of Remotely Sensed Data*, pp. 4b41-4b51, West Lafayette, Indiana.

Kawecki, T. J. 1993. Age and size at maturity in a patchy environment - fitness maximization versus evolutionary stability, *Oikos* 66 (2): 309-317.

Keeling, C. D., T. P. Whorf, M. Wahlen and J. van der Plicht. 1995. Interannual extremes in the rate of rise of atmospheric carbon dioxide since 1980, *Nature* 375 (6533): 666-670.

Kimes, D. S., S. B. Idso, P. J. Pinter, R. J. Reginato and R. D. Jackson. 1980. View angle effects in the radiometric measurement of plant canopy temperatures, *Remote Sensing of Environment* 10 : 273-284.

Kirschbaum, M. U. F., D. A. King, H. N. Comins, R. E. McMurtrie, B. E. Medlyn, S. Pongracic, D. Murty, H. Keith, R. J. Raison, P. K. Khanna and D. W. Sheriff. 1994. Modeling forest response to increasing CO_2 concentration under nutrient-limited conditions, *Plant, Cell and Environment* 17 : 1081-1099.

Koerper, G. E. and C. J. Richardson. 1980. Biomass and net annual primary production regressions for Populus grandidentata on three sites in northern lower Michigan, *Canadian Journal of Forest Research* 10 : 92-101.

K rner, C. and J. A. Arnone. 1992. Responses of elevated carbon dioxide in artifical tropical ecosystems, *Science* 257 : 1672-1675.

K rner, C., J. A. Scheel and H. Bauer. 1979. Maximum leaf diffusive conductance in vascular plants, *Photosynthetica* 13 : 45-82.

Korol, R. L., S. W. Running, K. S. Milner and E. R. Hunt. 1991. Testing a mechanistic carbon balance model against observed tree growth, *Canadian Journal of Forest Research* 21 : 1098-1105.

Kumar, M. and J. L. Monteith. 1982. Remote sensing of plant growth. Pages 133-144 *in* H. Smith, Ed., *Plants and the Daylight Spectrum*. London, Academic Press.

LaMarche, V. C. J., D. A. Graybill, H. C. Fritts and M. R. Rose. 1984. Increasing atmospheric carbon dioxide: tree ring evidence for growth enhancement in natural vegetation, *Science* 225: 1019-1021.

Lambers, H. and H. Poorter. 1992. Inherent variation in growth rate between higher plants: a search for physiological causes and ecological consequences, *Advances in Ecological Research* 23 : 187-261.

Lambers, H. and A. Rychter. 1989. The biochemical background of variation in respiration rate: respiratory pathways and chemical composition. Pages 199-225 *in* H. Lambers, M. Cambridge, H. Konings and T. L. Pons, Ed., *Causes and Consequences of Variation in Growth Rate and Productivity of Higher Plants*. The Hague, Academic Publishing.

Landsberg, J. J. 1986. *Physiological Ecology of Forest Production*, Applied botany: a series of monographs, Academic Press, London.

Landsberg, J. J., S. D. Prince, P. G. Jarvis, R. E. McMurtrie, R. Luxmore and B. E. Medlyn. 1996. Energy conversion and use in forests: the analysis of forest production in terms of utilisation efficiency ( ). Pages *in* H. L. Gholtz, K. Nakane and H. Shimoda, Ed., *Use of Remote Sensing in the Modeling of Forest Productivity at Scales from the Stand to the Globe*. New York, Kluwer Academic Publishers.

Landsberg, J. J. and L. L. Wright. 1989. Comparisons among Populus clones and intensive culture conditions using an energy conversion model, *Forest Ecology and Management* 27 (2): 129-147.

Larcher, W. 1995. *Physiological Plant Ecology*, Springer-Verlag, Berline, Heidelberg and New York.

Laurence, J. A., R. G. Amundson, A. L. Friend, E. J. Pell and P. J. Temple. 1994. Allocation of carbon in plants under stress: an analysis of the ROPIS experiments, *Journal of Environmental Quality* 23 : 412-417.

Lieth, H. 1975. Primary production of major vegetation units of the world, *in* H. Lieth and R. H. Whittaker, Ed., *Primary Production in the Biosphere*. New York, Springer-Verlag.

Lerdau, M. 1992. Future discounts and resource allocation in plants, *Functional Ecology* 6 : 371-375.

Levine, E. R., K. J. Ranson, J. A. Smith, D. L. Williams, R. G. Knox, H. H. Shugart, D. L. Urban and W. T. Lawrence. 1993. Forest ecosystem dynamics: linking forest succession, soil process and radiation models, *Ecological Modeling* 65 : 199-219.

Lieffers, V. J. and J. S. Campbell. 1984. Biomass and growth of Populus tremuloides in northeastern Alberta: estimates using hierarchy in tree size, *Canadian Journal of Forest Research* 14 : 610-616.

Lieffers, V. J. and S. E. MacDonald. 1990. Growth and foliar nutrient status of black spruce and tamarack in relation to depth of water table in some Alberta peatlands., *Canadian Journal of Forest Research* 20 : 805-809.

Linder, S. 1985. Potential and actual production in Australian forest stands. Pages 11-35 *in* J. J. Landsberg and W. Parsons, Ed., *Research for Forest Management*. Melbourne, CSIRO.

Loehle, C. 1988. Tree life history strategies: the role of defenses, *Canadian Journal of Forest Research* 18 : 209-222.

Lurin, B., W. Cramer, B. Moore III and S. I. Rasool. 1994. Global terrestrial net primary productivity, *International Geosphere Biosphere Program (IGBP) Global Change Newsletter* 19 : 6-8.

Maas, S. J. 1988. Use of remotely sensed information and agricultural crop growth models, *Ecological Modeling* 41 : 247-268.

MacArthur, R. H. and E. O. Wilson. 1967. *The Theory of Island Biogeography*, Princeton University Press, Princeton.

Mahendrappa, M. K. and P. O. Salonius. 1982. Nutrient dynamics and growth response in a fertilized black spruce stand, *Soil Science Society of America Journal* 46 : 127-133.

Matthews, E. 1983. Global vegetation and land use: new high resolution data bases for climate studies, *Climatology and Applied Meteorology* 22 : 474-487.

Maynard Smith, J. 1976. Evolution and the theory of games, *American Scientist* 64 : 41-45.

Maynard Smith, J. 1978. Optimization theory in evolution, *Annual Review Ecological Systematics* 9 : 31-56.

McCree, K. J. 1974. Equations for the rate of dark respiration of white clover and grain sorghum as functions of dry weight, photosynthetic rate and temperature, *Crop Science* 14 : 509-514.

McGuire, A. D., J. M. Melillo, L. A. Joyce, D. W. Kicklighter, A. L. Grace, B. Moore and C. J. Vorosmarty. 1992. Interactions between carbon and nitrogen dynamics in estimating net primary production for potential vegetation in North America, *Global Biogeochemical Cycles* 6 (2): 101-124.

McNaughton, K. G. and P. G. Jarvis. 1991. Effects of spatial scale on stomatal control of transpiration, *Agricultural and Forest Meteorology* 54 : 279-301.

Melillo, J. M., A. D. McGuire, D. W. Kicklighter, B. Moore, C. J. Vorosmarty and A. L. Schloss. 1993. Global climate change and terrestrial net primary production, *Nature* 363 : 234-240.

Monson, R. K., E.-D. Schulze, M. Freund and H. Heilmeier. 1994. The influence of nitrogen availability on carbon and nitrogen storage in biennial *Cirsium vulgare (Savi)* II. The cost of nitrogen storage, *Plant, Cell and Environment* 17 : 1133-1141.

Monteith, J. L. 1972. Solar radiation and productivity in tropical ecosystems, *Journal of Applied Ecology* 9 : 747-766.

Monteith, J. L. 1977. Climate and the efficiency of crop production in Britain, *Philosophical Transactions of the Royal Society of London (B)*, 281 : 277-294.

Mooney, H. A., W. E. Winner and E. J. Pell. 1991. *Response of plants to multiple stresses*, Physiological Ecology, Academic Press, San Diego.

Moore, B., R. D. Boone, J. E. Hobbie, R. A. Houghton, J. M. Mellilo, B. J. Peterson, G. R. Shaver, C. J. Vorosmarty and G. M. Woodwell. 1981. A simple model for analysis of the role of terrestrial ecosystems in the global carbon budget. Pages 365-285 *in* B. Bolin, Ed., *Carbon Cycle Modeling*. Chichester, England, John Wiley & Sons.

Munson, A. D. and V. R. Timmer. 1990. Site-specific growth and nutrition of planted *Picea mariana* in the Ontario Clay Belt: III. Biomass and nutrient allocation., *Canadian Journal of Forest Research* 20 (8): 1165-1180.

Myneni, R. B., G. Asrar and F. G. Hall. 1992a. A three dimensional radiative transfer method for optical remote sensing of vegetated land surfaces, *Remote Sensing of Environment* 40 (2):105-116.

Myneni, R. B., G. Asrar, D. Tanre and B. J. Choudhury. 1992b. Remote sensing of solar radiation absorbed and reflected by vegetated land surfaces, *IEEE Transactions on Geoscience and Remote Sensing* 30 (2): 302-314.

116

Nordin, J. O. and D. F. Grigal. 1976. Vegetation, site, and fire relationships within the area of the Little Sioux Fire, northeastern Minnesota, *Canadian Journal of Forest Research* 6 : 78-85.

Norman, J. M. and P. G. Jarvis. 1974. Photosynthesis in Sitka spruce (*Picea sitchensis* (Bong) Carr). III. Measurements of canopy structure and interception of radiation, *Journal of Applied Ecology* 11 : 375-398.

Oliver, C. D. and B. C. Larson. 1990. *Forest stand dynamics*, McGraw-Hill, Inc, New York.

Overpeck, J. T., P. J. Bartlein and T. Webb. 1991. Potential magnitude of future vegetation change in eastern North America: comparisons with the past, *Science* 254 : 692-695.

Parker, G. A. and J. M. Smith. 1990. Optimality theory in evolutionary biology, *Nature* 348 : 27-33.

Parton, W. J., J. W. B. Stewart and C. V. Cole. 1988. Dynamics of C, N, P, and S in grassland soils: a model, *Biogeochemistry* 5 : 109-131.

Partridge, L. and P. H. Harvey. 1988. The ecological context of life history evolution, *Science* 241 : 1449-1455.

Pastor, J. and J. G. Bockheim. 1984. Distribution and cycling of nutrients in an aspen-mixed-hardwood-spodosol ecosystem in northern Wisconsin, *Ecology* 65 (2): 339-353.

Pastor, J., R. H. Gardner, V. H. Dale and W. M. Post. 1987. Successional changes in nitrogen availability as a potential factor contributing to spruce declines in boreal North America, *Canadian Journal of Forest Research* 17 : 1394-1400.

Pastor, J. and W. M. Post. 1988. Response of northern forests to CO_2-induced climate change., *Nature* 334 (6177): 55-58.

Pauley, S. S. and T. O. Perry. 1954. Ecotypic variation of the photosynthetic response in *Populus*, *Journal of the Arnold Arboretum* 35 : 167-188.

Pearman, G. I. and P. Hyson. 1981. The annual variation of atmospheric CO_2 concentration observed in the northern hemisphere., *Journal of Geophysical Research* 86 (C10): 9839-9843.

Penning de Vries, F. W. T. 1975. The cost of maintenance processes in plant cells, *Annals of Botany* 39 : 77-92.

Perrin, N. and R. M. Sibley. 1993. Dynamic models of energy allocation and investment, *Annual Review of Ecology and Systematics* 24 : 379-410.

Poorter, H., A. Van der Werf, O. K. Atkin and H. Lambers. 1991. Respiratory energy requirements of roots vary with the potential growth rate of a plant species, *Physiologia Plantarum* 83 : 469-475.

Post, W. M., T. Peng, W. R. Emmanuel, A. W. King, V. H. Dale and D. L. DeAngelis. 1990. The global carbon cycle, *American Scientist* 78 : 310-326.

Potter, C. S., J. T. Randerson, C. B. Field, P. A. Matson, P. M. Vitousek, H. A. Mooney and S. A. Klooster. 1993. Terrestrial ecosystem production: A process model based on global satellite and surface data, *Global Biogeochemical Cycles* 7 (4): 811-841.

Prentice, I. C. 1986. Vegetation response to past climatic variation, *Vegetatio* 67 : 131-141.

Prince, S. D. 1991a. Satellite remote sensing of primary production: comparison of results for Sahelian grasslands 1981-1988, *International Journal of Remote Sensing* 12 : 1301-1330.

Prince, S. D. 1991b. A model of regional primary production for use with coarse-resolution satellite data, *International Journal of Remote Sensing* 12 (6): 1313-1330.

Prince, S. D. and S. J. Goward. 1995. Global primary production: a remote sensing approach, *Journal of Biogeography* 22.

Prince, S. D., C. O. Justice and B. Moore. 1994. Monitoring and modeling of terrestrial net and gross primary production, *International Geosphere Biosphere Program (IGBP) Data and Information System (DIS) Global Analysis, Interpretation and Modeling (GAIM)*, Working Paper #1.

Prince, S. D. and C. J. Tucker. 1986. Satellite remote sensing of rangelands in Botswana II: NOAA AVHRR and herbaceous vegetation, *International Journal of Remote Sensing* 7 (11): 1555-1570.

Pugnaire, F. I. and F. S. Chapin III. 1992. Environmental and physiological factors governing nutrient resorption efficiency in barley, *Oecologia* 90 : 120-126.

Raich, J. W. and K. J. Nadelhoffer. 1989. Belowground carbon allocation in forest ecosystems: global trends, *Ecology* 70 (5): 1346-1354.

Raich, J. W., E. B. Rastetter, J. M. Melillo, D. W. Kickkighter, P. A. Steudler, B. J. Peterson, A. L. Grace, B. M. III and C. J. Vorosmarty. 1991. Potential net primary productivity in South America: application of a global model, *Ecological Applications* 1 (4): 399-429.

Ramanathan, V. 1988. The greenhouse theory of climate change: a test by inadvertent global experiment, *Science* 240 : 293-299.

Rauner, J. L. 1976. Deciduous Forests. Pages 241-264 *in* J. L. Monteith, Ed., *Vegetation and the Atmosphere*. New York, Academic Press.

Raven, J. A. 1986. Evolution of plant life forms. Pages 421-492 *in* T. J. Givnish, Ed., *On the economy of plant form and function*. London, England, Cambridge University Press.

Raynaud, D., J. Jouzel, J. M. Barnola, J. Chappellaz, R. J. Delmas and C. Lorius. 1993. The ice record of greenhouse gases, *Science* 259 : 926-934.

Reich, P. B., M. B. Walters and D. S. Ellsworth. 1992. Leaf life-span in relation to leaf, plant, and stand characterisitcs among diverse ecosystems, *Ecological Monographs* 62 (3): 365-392.

Richardson, A. J. and C. L. Wiegand. 1989. Canopy leaf display effects on absorbed, transmitted, and reflected solar radiation, *Remote Sensing of Environment* 29 : 15-24.

Ricklefs, R. E. 1991. Structures and transformations of life histories, *Functional Ecology* 5 : 174-183.

Rosenzweig, M. L. 1968. Net primary productivity of terrestrial communities: prediction from climatological data, *American Naturalist* 102 (923): 67-74.

Roujean, J.-L. and F.-M. Breon. 1995. Estimating PAR absorbed by vegetation from bidirectional reflectance measurements, *Remote Sensing of Environment* 51 : 375-384.

Rowe, J. S. 1972. *Forest regions of Canada*, Forestry Service, Environment Canada, Publication #3100.

Ruark, G. A. and J. G. Bockheim. 1987. Below-ground biomass of 10-, 20-, and 32-year-old Populus tremuloides in Wisconsin, *Pedobiologia* 30 : 207-217.

Ruark, G. A. and J. G. Bockheim. 1988. Biomass, net primary production, and nutrient distribution for an age sequence of *Populus tremuloidies* ecosystems., *Canadian Journal of Forest Research* 18 : 435-443.

Ruimy, A., G. Dedieu and B. Saugier. 1994. Methodology for the estimation of terrestrial net primary production from remotely sensed data, *Journal of Geophysical Research* 99 (D3): 5263-5283.

Running, S. W. and J. C. Couglan. 1988. A general model of forest ecosystem processes for regional applications: I. Hydrological balance, canopy gas exchange and primary production processes, *Ecological Modeling* 42 : 125-154.

Running, S. W. and S. T. Gower. 1991. FOREST-BGC, a general model of forest ecosystem processes for regional applications. II. Dynamic carbon allocation and nitrogen budgets, *Tree Physiology* 9 : 147-160.

Running, S. W. and E. R. Hunt. 1993. Generalization of a forest ecosystem process model for other biomes, BIOME-BGC, and an application for global-scale models. Pages 141-158 *in* J. R. Ehleringer and C. B. Field, Ed., *Scaling Physiological Processes*. San Diego California, Academic Press, Inc.

Running, S. W. and R. R. Nemani. 1987. Relating seasonal patterns of the AVHRR vegetation index to simulated photosynthesis and transpiration of forests in different climates, *Remote Sensing of Environment* 24 : 347-367.

Running, S. W., R. R. Nemani, D. L. Peterson, L. E. Band, D. F. Potts, L. L. Pierce and M. A. Spanner. 1989. Mapping regional forest evapotranspiration and photosynthesis by coupling satellite data with ecosystem simulation, *Ecology* 70 (4): 1090-1101.

Runyon, J., R. H. Waring, S. N. Goward and J. M. Welles. 1994. Environmental limits on net primary production and light use efficiency across the Oregon transect, *Ecological Applications* 4 (2): 226-238.

Ryan, M. G. 1990. Growth and maintenance respiration in stems of *Pinus contorta* and *Picea engelmannii*, *Canadian Journal of Forest Research* 20 : 48-57.

Ryan, M. G. 1995. Foliar maintenance respiration of supalpine and boreal trees and shrubs in relation to nitrogen content, *Plant, Cell and Environment* 18 : 765-772.

Ryan, M. G., R. M. Hubbard, D. A. Clark and R. L. Sanford. 1994. Woody-tissue respiration for *Simarouba amara* and *Minquartia guianensis*, two tropical wet forest trees with different growth habits, *Oecologia* 100 : 213-220.

Ryan, M. G. and R. H. Waring. 1992. Maintenance respiration and stand development in a subalpine lodgepole pine forest, *Ecology* 73 : 2100-2108.

Saldarriaga, J. G. and R. J. Luxmoore. 1991. Solar energy conversion efficiencies during succession of a tropical rain forest in Amazonia, *Journal of Tropical Ecology* 7 : 233-242.

Santantonio, D., R. K. Herman and W. S. Overton. 1977. Root biomass studies in forest ecosystems, *Pedobiologia* 17 : 1-31.

Sato, N., P. J. Sellers, D. A. Randall, E. K. Schneider, J. Shukla, J. L. Kinter, Y.-T. Hou and E. Albertazzi. 1989. Effects of implementing the Simple Biosphere (SiB) model in a general circulation model, *Journal of Atmospheric Science* 46 (18): 2757-2782.

Schmid, B. and F. A. Bazzaz. 1994. Crown construction, leaf dynamics and carbon gain in two perennials with contrasting architecture, *Ecological Monographs* 64 (2): 177-203.

Schneider, S. H. 1989. The greenhouse effect: science and policy, *Science* 243 : 771-780.

Sellers, P. J. 1985. Canopy reflectance, photosynthesis and transpiration, *International Journal of Remote Sensing* 6 : 1335-1372.

Sellers, P. J. 1987. Canopy reflectance, photosynthesis and transpiration. II. The role of biophysics in the linearity of their interdependence, *Remote Sensing of Environment* 21 : 143-183.

Sellers, P. J., J. A. Berry, G. J. Collatz, C. B. Field and F. G. Hall. 1992. Canopy reflectance, photosynthesis and transpiration III: A reanalysis using improved leaf models and a new canopy integration scheme, *Remote Sensing of Environment* 42 : 187-216.

Sellers, P. J., Y. Mintz, Y. C. Sud and A. Dalcher. 1986. A simple biosphere model (SiB) for use with general circulation models, *Journal of Atmospheric Science* 43 : 505-531.

Shipley, B. and R. H. Peters. 1990. A test of the Tilman model of plant strategies: relative growth rate and biomass partitioning, *American Naturalist* 136 : 139-153.

Shugart, H. H. and D. C. West. 1977. Development of an Appalachian deciduous forest succession model and its application to assessment of the impact of the chestnut blight, *Journal of Environment and Man* 5 : 161-179.

Shukla, J., C. Nobre and P. Sellers. 1990. Amazon deforestation and climate change, *Science* 247 : 1322-1325.

Sibley, R. M. 1989. What evolution maximizes, *Functional Ecology* 3 : 129-135.

Sibley, R. M. 1991. The life-history approach to physiological ecology, *Functional Ecology* 5 : 184-191.

Silvertown, J., M. Franco and K. McConway. 1992. A demographic interpretation of Grime s triangle, *Functional Ecology* 6 : 130-136.

Sobrado, M. A. 1991. Cost-benefit relationships in deciduous and evergreen leaves of tropical dry forest species, *Functional Ecology* 5 : 608-616.

Sobrado, M. A. 1994. Leaf age effects on photosynthetic rate, transpiration rate and nitrogen content in a tropical dry forest, *Physiologia Plantarum* 90 : 210-215.

Stearns, S. C. 1976. Life history tactics: a review of the ideas, *Quarterly Review of Biology* 51 (1): 3-47.

Stearns, S. C. 1989. Trade-offs in life-history evolution, *Functional Ecology* 3 : 259-268.

Steven, M. D., P. V. Biscoe and K. W. Jaggard. 1983. Estimation of sugar beet productivity from reflection in the red and infrared spectral bands, *International Journal of Remote Sensing* 4 (2): 325-334.

Strahler, A. H. and D. L. B. Jupp. 1991. Geometric-optical modeling of forests as remotely-sensed scenes composed of three-dimensional, discrete objects. Pages 415-440 *in* R. Myneni and J. Ross, Ed., *Photon-Vegetation Interactions*. New York, Springer-Verlag.

Sundquist, E. T. 1993. The global carbon dioxide budget, *Science* 259 : 934-941.

Tans, P. P., I. Y. Fung and T. Takahashi. 1990. Observational constraints on the global atmospheric CO_2 budget, *Science* 247 : 1431-1438.

Thompson, D. J. 1995. The seasons, global temperature, and precession, *Science* 268 : 59-68.

Tilman, D. 1988. *Plant strategies and the dynamics and structure of plant communities*, Princeton University Press, Princeton, New Jersey.

Tilman, D. 1991. Relative growth rates and plant allocation patterns, *American Naturalist* 138 : 1269-1275.

Tucker, C. J., I. Y. Fung, C. D. Keeling and R. H. Gammon. 1986. Relationship between atmospheric CO_2 variations and a satellite-derived vegetation index, *Nature* 319 : 195-199.

Tucker, C. J., C. L. Vanpraet, E. Boerwinkel and A. Gaston. 1983. Satellite remote sensing of total dry matter accumulation in the Senegalese Sahel, *Remote Sensing of Environment* 13 : 461-474.

Valentini, R., J. A. Gamon and C. B. Field. 1995. Ecosystem gas exchange in a California grassland: Seasonal patterns and implications for scaling, *Ecology* 76 (6): 1940-1952.

Van Cleve, K., L. Oliver, R. Schlentner, L. Viereck and C. T. Dyrness. 1983. Productivity and nutrient cycling in taiga forest ecosystems, *Canadian Journal of Forest Research* 13 : 747-766.

Van Cleve, K. and L. K. Oliver. 1982. Growth response of postfire quaking aspen to N, P, and K fertilization, *Canadian Journal of Forest Research* 12 : 160-165.

van der Werf, A., A. J. Visser, F. Schieving and H. Lambers. 1993. Evidence for optimal partitioning of biomass and nitrogen availabilities for a fast- and slow-growing species, *Functional Ecology* 7 : 63-74.

Verma, S. B., D. D. Baldocchi, D. E. Anderson, D. R. Matt and R. J. Clement. 1986. Eddy fluxes of CO_2, water vapor, and sensible heat over a deciduous forest, *Boundary Layer Meteorology* 36 : 71-91.

Viereck, L. A., C. T. Dyrness, K. V. Cleve and M. J. Foote. 1983. Vegetation, soils, and forest productivity in selected forest types in interior Alaska, *Canadian Journal of Forest Research* 13 : 703-720.

Vitousek, P. 1982. Nutrient cycling and nutrient use efficiency, *American Naturalist* 119 (4): 553-572.

Vorosmarty, C. J., B. M. III, A. L. Grace, M. P. Gildea, J. M. Melillo, B. J. Peterson, E. B. Rastetter and P. A. Steudler. 1989. Continental scale models of water balance and fluvial transport: an application to South America, *Global Biogeochemical Cycles* 3 (3): 241-265.

Walters, M. B. and C. B. Field. 1987. Photosynthetic light acclimation in two rainforest Piper species with different ecological amplitudes, *Oecologia* 72 : 449-456.

Waring, R. H. 1983. Estimating forest growth and efficiency in relation to canopy leaf area, *Advances in Ecological Research* 13 : 327-354.

Waring, R. H., B. E. Law, M. L. Goulden, S. L. Bassow, R. W. McCreight, S. C. Wofsy and F. A. Bazzaz. 1995. Scaling gross ecosystem production at Harvard Forest with remote sensing: a comparison of estimates from a constrained quantum-use efficiency model and eddy correlation, *Plant Cell Environment* 18 : 1201-1213.

Whittaker, R. H. and G. E. Likens. 1973. Carbon in the biota. Pages 281-300 *in* G. M. Woodwell and E. V. Pecan, Ed., *Carbon and the Biosphere*. Springfield Virginia, National Technical Information Service.

Williams, K., C. B. Field and H. A. Mooney. 1989. Relationships among leaf construction cost, leaf longevity, and light environment in rain-forest plants of the genus *Piper*, *American Naturalist* 133 (2): 198-211.

Williams, W. E., K. Garbutt, F. A. Bazzaz and P. M. Vitousek. 1986. The response of plants to elevated CO_2, *Oecologia* 69 : 454-459.

Woods, K. D. 1988. Estimation of above-ground net primary production of trembling aspen and lowland black spruce in the Superior National Forest, Minnesota. Unpublished memo.

Woods, K. D., A. H. Fieveson and D. B. Botkin. 1991. Statistical error analysis for biomass density and leaf area index estimation, *Canadian Journal of Forest Research* 21 : 974-989.

Woodward, F. I., G. B. Thompson and I. F. McKee. 1991. The effects of elevated concentrations of carbon dioxide on individual plants, populations, communities and ecosystems, *Annals of Botany* 67 : 23-38.

Wright, H. E. and W. A. Watts. 1969. *Glacial and vegetational history of northeastern Minnesota*, University of Minnesota, Geologic Survey Special Publication SP-11, 59 pp.

Yocum, C. S., L. H. Allen and E. R. Lemon. 1964. Photosynthesis under field conditions. VI. Solar radiation balance and photosynthetic efficiency, *Agronomy Journal* 56 : 249-253.